To Elizabeth and Fam[ily]
with all best wishes
Crispin, September 2012

Managing Engineering Design

Springer
*London
New York
Berlin
Heidelberg
Hong Kong
Milan
Paris
Tokyo*

Crispin Hales and Shayne Gooch

Managing Engineering Design

Second Edition

With 76 Figures

 Springer

Crispin Hales, Ph.D, C.Eng, F.I.Mech.E
Triodyne Inc., 666 Dundee Road, Northbrook, IL 60062-2702, USA

Shayne Gooch, Ph.D
University of Canterbury, Christchurch, New Zealand

British Library Cataloguing in Publication Data
Hales, Crispin
  Managing engineering design. – 2nd ed.
  1. Engineering design – Management
  I. Title II. Gooch, Shayne
  620′.0042′068
  ISBN 1-85233-803-2

Library of Congress Cataloging-in-Publication Data
Hales, Crispin, 1945–
  Managing engineering design / Crispin Hales, Shayne Gooch. – 2nd ed.
    p. cm.
  ISBN 1-85233-803-2 (alk. paper)
  1. Engineering design – Management. I. Gooch, Shayne. II. Title.
  TA174.H23 2004
  620′.0042 – dc22
                                                            2004042560

Apart from any fair dealing for the purposes of research or private study, or criticism or review, as permitted under the Copyright, Designs and Patents Act 1988, this publication may only be reproduced, stored or transmitted, in any form or by any means, with the prior permission in writing of the publishers, or in the case of reprographic reproduction in accordance with the terms of licences issued by the Copyright Licensing Agency. Enquiries concerning reproduction outside those terms should be sent to the publishers.

ISBN 1-85233-803-2 Springer-Verlag London Berlin Heidelberg
Springer-Verlag is part of Springer Science+Business Media
springeronline.com

First edition published by Longman Scientific & Technical, ISBN 0-582-03933-9

© Springer-Verlag London Limited 2004

The use of registered names, trademarks, etc. in this publication does not imply, even in the absence of a specific statement, that such names are exempt from the relevant laws and regulations and therefore free for general use.

The publisher makes no representation, express or implied, with regard to the accuracy of the information contained in this book and cannot accept any legal responsibility or liability for any errors or omissions that may be made.

Typesetting: SNP Best-set Typesetter Ltd., Hong Kong
Printed and bound in the United States of America
69/3830-543210 Printed on acid-free paper SPIN 10836592

Preface to Second Edition

Engineering design concerns us all. It affects our everyday lives and increasingly affects the future of life on this planet. The time has gone when design engineers were told what was required and did their best to come up with something that worked. Competition is fierce, markets are international, and the consequences of poor design are felt globally. There is strong pressure for shorter project timescales and higher quality design at lower cost. Designs must work, they must be culturally and politically acceptable, and they must be safe, reliable and environmentally sound. A failure in any one of these aspects can result in bankruptcy or disaster, and to avoid such debilitating situations the design engineer needs the genuine support of all parties involved: management; marketing; manufacturing; customers and users. It is no longer acceptable for design engineers to work in isolation from everyone else, and it is no longer acceptable for everyone else to plead ignorance of the design engineer's work. We are all involved with design and we all have a responsibility to make sure that design is done in the best possible way.

So what is the role of a design engineer? A design engineer is presented with a technical problem or need, and the ultimate aim is the conversion of this into the information from which something can be manufactured at high enough quality and low enough cost to overcome the problem or to meet the need. This may sound simple, but in fact so many factors influence the situation that it is often difficult for one person to understand the problem fully, let alone produce solutions that meet everyone's expectations. Design is a team activity. Communication and information exchange are critical.

The manager responsible for engineering design must understand the problem or need in its overall context, must be able to build up a strong working team within that context, and must be able to steer the project through the design process to the point where manufacture is in progress. From then on there is a reduced, but important, responsibility to monitor the performance of the design in practice and ensure that it continues to satisfy customer and user needs throughout its life. Feedback of performance information is essential for future development.

It is not possible to cope with all the issues using an "inventor" approach to design, neither is it necessary. We now know enough about the design process, the working of teams, and the communication of information to be able to tackle a design project in a systematic and confident manner. The rapid advances in computing and communication technology during the decade since the first edition of this book was published have enabled the implementation of approaches to engineering design that have long since been developed, but which have lacked the practical means for delivery. The use of Web-based design aids and geographically dispersed design teams has become a reality, and it is now possible to work in ways that would not have been considered a few years ago. There is little excuse for poor design.

This book brings together some guidelines for the management of engineering design projects within a Web-based framework that encourages a systematic approach to design. It is based on the results of experience in industry combined with the results of academic design research, and it includes a unique series of checklists and work sheets for direct application on projects. The checklists pose a structured set of questions for the design manager to use during each phase of the design process, and the work sheets provide a means for summarizing the project status at any particular time. They can be used before or during design review meetings to highlight action items and, when collected together, they form a historical record of project progress.

Many people offered suggestions and encouragement during the original development of this book, but there are two in particular whose invaluable help must be especially acknowledged. Firstly there was Ken Wallace, Professor of Engineering Design at Cambridge University in the UK. It was Ken who, in the early 1980s, translated the systematic engineering design approach as presented by Professor Pahl and Professor Beitz in Germany. This has become our cornerstone for both design teaching and design practice. Secondly there was Tom Zabinski, of the Graphics Communication Department at Triodyne Inc. in the USA. It was Tom who spent many long hours making the complicated diagrams more understandable to the reader and laying out the checklists and work sheets in a practicable form, which ultimately could be converted to a Web-based system.

We selected the Life chair, designed by Formway in New Zealand but manufactured and marketed by Knoll in the USA, to provide a working example of successful product design. The help of both companies is gratefully acknowledged, and in particular the enthusiastic involvement of Jon Prince, Design Team Leader for the chair project. We would also like to thank Katherine Vyver and Andrea Roberts from Formway Design for providing final images and proofreading sections of the text. A detailed review of the project was undertaken, and its history was reconstructed chronologically by questioning according to the checklists. The checklists were used in the same order as presented in this book, and the corresponding work sheets were filled out as the reconstruction took place. The result was a set of completed work sheets that have been used as examples in sequence throughout this book.

The Triodyne Safety Information Center provided help with the research on standards and codes, and a set of five reference papers written by Triodyne staff laid a foundation for the text. The help of Marna Sanders is acknowledged in bringing together current information with regard to the sourcing of relevant standards and codes for engineering design. Using the Triodyne facilities, she was able to compile a useful bibliography on standardization and a comprehensive international list of Website addresses for obtaining standards and codes.

There are now many helpful books available on the management of projects, on the engineering design process, product design, concurrent engineering and specific design techniques. However, when it actually comes to managing an engineering design project within a company, circumstances often make it difficult to apply all but the simplest techniques. There are some subtle day-to-day issues that are time consuming, frustrating and difficult to handle, yet which have not been addressed adequately in the literature. They are sometimes referred to as the "hidden costs of design." What has been attempted here is to present a systematic and practical approach to handling such issues by considering first the context within which the design work will take place, then the nature of the project, the design team and the available tools, and then each phase of the design process itself. As the book is intended to complement texts on project management, design methods, and specific areas of design, references are given and further reading suggestions are provided in the Bibliography. The underlying idea is to help the design manager operate effectively and efficiently by integrating multidisciplinary viewpoints and coordinating the design process at every level within a company. If it helps to improve the quality of our engineering design for the future then it will have done its job.

Crispin Hales and Shayne Gooch
Northbrook, IL, USA and
Christchurch, New Zealand
April 2004

Contents

Introduction		1
0.1	Terminology	2
0.2	Examples	3

PART 1 The Context 9

1 Ways of Thinking about Engineering Design 11

    1.1 Disasters and Failures 11
    1.2 Engineering Excellence 15
    1.3 New Innovations 15
    1.4 Improving Engineering Design 17
    1.5 Systematic Approaches to Engineering Design 18
    1.6 Systematic Design in Practice 20
    1.7 *Tips for Management* 22

2 The Project Context 23

    2.1 Engineering Projects 23
    2.2 Engineering Design in the Project Context 23
    2.3 The Effect of Influences 29
    2.4 Influences at the Macroeconomic Level 31
    2.5 Influences at the Microeconomic Level 34
    2.6 Influences at the Corporate Level 39
    2.7 Design Context Checklist and Work Sheet 43
    2.8 *Tips for Management* 51

PART 2 Task, Team and Tools 53

3 Profiling the Project 55

    3.1 Influences at the Project Level 55
    3.2 Design Task 55
    3.3 Design Team 58
    3.4 Design Tools and Techniques 62
    3.5 Design Team Output 67

	3.6	*Project Profile Checklist and Work Sheet*	76
	3.7	*Tips for Management*	80
4	Managing the Design Team	83
	4.1	Influences at the Personal Level	83
	4.2	Knowledge, Skills, and Attitude	84
	4.3	Motivation	85
	4.4	Relationships	87
	4.5	Personal Output	87
	4.6	*Personnel Profile Checklist and Work Sheet*	88
	4.7	*Tips for Management*	92

PART 3 The Project … 93

5	Project Proposal: Getting the Job	95
	5.1	Proposals and Briefs	95
	5.2	Preparing a Proposal	96
	5.3	Negotiations	100
	5.4	Debriefing	101
	5.5	*Project Proposal Checklist and Work Sheet*	101
	5.6	*Tips for Management*	102
6	Design Specification: Clarification of the Task	107
	6.1	Problem Statement and Design Specification	107
	6.2	Defining the Problem	108
	6.3	Project Planning	109
	6.4	Demands and Wishes	110
	6.5	Design Specification	111
	6.6	*Design Specification Checklist and Work Sheet*	114
	6.7	*Tips for Management*	115
7	Feasible Concept: Conceptual Design	119
	7.1	Divergent and Convergent Thinking	119
	7.2	Generating Ideas	122
	7.3	Selecting and Evaluating Concepts	126
	7.4	Estimating Costs	127
	7.5	Presenting the Final Concept	130
	7.6	*Conceptual Design Checklist and Work Sheet*	133
	7.7	*Tips for Management*	140
8	Developed Concept: Embodiment Design	141
	8.1	Abstract Concept to Developed Design	141
	8.2	Overall Guidelines for Embodiment Design	149

	8.3	Specific Guidelines for Embodiment Design	154
	8.4	General Guidelines for Embodiment Design	159
	8.5	*Embodiment Design Checklist and Work Sheet*	169
	8.6	*Tips for Management*	176
9	Final Design: Detail Design for Manufacture		177
	9.1	The Importance of Detail Design	177
	9.2	The Design Manager and Detail Design	178
	9.3	Quality Assurance	178
	9.4	Interaction of Shape, Materials, and Manufacture	180
	9.5	Manufacturing Drawings and Information	189
	9.6	Standard Components	190
	9.7	Assembly	190
	9.8	Testing and Commissioning	191
	9.9	*Detail Design Checklist and Work Sheet*	197
	9.10	*Tips for Management*	204
10	Users and Customers: Design Feedback		205
	10.1	Expectations	205
	10.2	Use and Abuse	206
	10.3	Maintenance	208
	10.4	Litigation	208
	10.5	*Design Quality Assessment Work Sheet*	211
	10.6	*Tips for Management*	212
11	Standards and Codes		217
	11.1	General Issues	217
	11.2	Basic Definitions	218
	11.3	Safety Standards	220
	11.4	Some Reference Articles on Safety Standards	221
	11.5	Some Reference Articles on International Standards	221
	11.6	ISO 9000 International Standards for Quality Management	222
	11.7	National Standards for Engineering Design Management	222
	11.8	*Tips for Management*	223
	11.9	*Contact Information and URLs for Standards and Codes*	223
12	Engineering Design Process: Review and Analysis		233
	12.1	Summary	233
	12.2	Forensic Analysis of Engineering Design Issues	235
	12.3	Analysis of the Engineering Design Process	236

xii   Contents

References 241
Bibliography 243
Index 247

CD-ROM: Managing Engineering Design Tools

Information Files:
   Indexed Bibliography
   Indexed Links to Standards Organizations
   Links to MED Interactive Web Site

Working Files:
   Printable Work Sheets
   Web Based Work Sheets
   Interactive Web Based Work Sheets

Introduction

0.1 Terminology
0.2 Examples

To convert an idea or a need into the information from which a new product or system can be made requires a transformation from vague concepts to defined objects, from abstract thoughts to the solution of detailed problems. It is through the engineering design process that this transformation takes place. Successful management of this process boils down to the effective handling of three issues:

- *Activities* of the design team.
- *Output* from the design team.
- *Influences* on the design team.

The activities of the design team must be guided and monitored for performance. The design output must be continually reviewed and assessed for quality. The effect of influencing factors must be forecast, monitored, and controlled where possible. Genuine management attention to these issues is crucial to the development of high-quality and cost-competitive products. This means clear understanding and observant monitoring of the design process, awareness of potential problems, and skillful management of complex design situations. From the design management point of view, the ultimate goal is to produce the highest quality product that meets the user's expectations for the lowest cost in the shortest time.

Part 1 of this book is concerned with the overall *context* within which the engineering design process takes place. It offers a way of mapping the context for specific design situations so as to identify key influences and then to take advantage of those that are positive, while minimizing the effect of negative influences. Part 2 of this book is concerned with matching the *design team* capabilities and its activities to the specific design task in order to maximize the productivity and quality of the team output. Part 3 of this book presents a Web-based and structured approach to managing engineering *design projects*, which can be adapted readily to suit personal preferences within a particular company or situation. Each part of the book includes *checklists* and *work sheets* in a Web-based format, for use by the manager to understand better the status of a project at any particular time and thereby to determine the best course of action. The checklists provide questions to be raised by the design manager, and the work sheets record the current project status for use at project meetings and as a permanent record for future reference.

This book is not intended as a technical engineering design text, nor as a business guide, but presents a way of looking at what design work involves, how specific aspects can be monitored in practice, and how the results may be used to improve the design management within a company. The approach is necessarily a hybrid of the quantitative and the qualitative. A particular difficulty with the management of engineering design is that the critical issues are wide ranging within the spectrum from "hard" to "soft": from the dimensional tolerance on a component, to the user's satisfaction with a product in service. Another difficulty is that the critical issues must be considered at different levels and from different points of view. A key management skill in engineering design is to have a good grasp of the overall picture, but with the ability to window in rapidly on the tiniest of technical details and understand the effect that such a seemingly insignificant detail might have on the overall project. Many engineering disasters, such as the failure of the solid rocket booster on Space Shuttle Challenger, have come about through a lack of management skill in this area. The checklists and work sheets are provided to help in this "windowing" process, and each chapter concludes with a set of useful *tips for management*. It becomes clear that simple sets of questions, based on fundamental design principles, and asked at the appropriate time by a manager with adequate technical understanding, can highlight design weaknesses long before the road to disaster is inevitable.

0.1 Terminology

Many engineering design terms vary in meaning according to discipline, context, and interpretation. To overcome the problem in this book, simple terms with generally accepted meanings are used wherever possible, and the number of terms used has been minimized. When "design" is used it refers to "engineering design" unless stated otherwise, and both terms may refer occasionally to the "field" of enquiry or of practice. "Design engineer," "engineering designer," and "designer" have been treated as synonymous, but "design engineer" is preferred. General terms such as "specification" are used with qualifiers to clarify their meaning in a particular context; for example: "design specification"; "materials specification"; and "test specification." Sometimes, a further qualifier might be added for more precision, such as in "product design specification" or "system design specification."

There is often great discussion over what exactly is meant by the term "engineering design". For the purposes of this book, the following definition will be used:

Engineering design is the process of converting an idea or market need into the detailed information from which a product or technical system can be produced.

0.2 Examples

Practical examples from personal experience are used to clarify the presentation. Some are mentioned in passing to illustrate a certain point, some are described in more detail to highlight important issues, and two have been used as reference examples throughout the book. The two reference examples are purposely quite different from each other: one involves a complex one-off high-pressure test system and the other involves a mass-produced office chair. The first was a 3-year project in the UK, recorded and analyzed in great detail as part of one author's doctoral research at Cambridge University. The other was a 5-year collaborative project carried out by well-known companies in the USA and in New Zealand. This resulted in a patented office chair, called "Life," which has been recognized for its innovative ergonomic and environmental features. The Life chair is a mass-produced product and is marketed worldwide. Each of these two examples is described briefly below.

0.2.1 Gasifier Test Rig

The gasifier test-rig project involved the design of a high-pressure and high-temperature system for evaluation of materials in a simulated slagging coal-gasifier environment (Figure 0.1). The company considered the design task to be challenging, because of the extreme test conditions. A slagging coal-gasifier converts coal to natural gas at such high temperatures that the ash melts to form a protective but corrosive slag, similar to that produced as a by-product in a blast furnace. Very few metal alloys are capable of withstanding the combination of temperature, pressure, and corrosive environment inherent in the slagging coal-gasification process; hence, the need for materials testing under laboratory conditions. In this particular materials test facility, or test rig, automatic control of temperature, gas flows, liquid flows and coal flows at high pressure for continuous periods of up to 1000 h was required. The main difficulty, and the novel feature of the proposed system, lay in the handling of flowing coal on such a small scale under such extremes of pressure and temperature. Although the need for such equipment had been identified by the company, the requirements had not been formally established and the ideas were vague; so, in engineering design terms, the problem was "ill-defined." The only unusual aspects of the project itself from the company's viewpoint were: that a systematic approach would be applied to the complete design effort; and that every activity related to the project would be recorded in detail for analysis. It was never intended that this particular project should be completed in the minimum possible time, but rather that the design effort should be integrated with work on other projects and should extend over 2 years to fit in with annual budget constraints.

The project proposal, submitted to the materials research group within one division of the company, was accepted and the design work started as soon as a contract was signed. The first activity was to clarify the design task, by

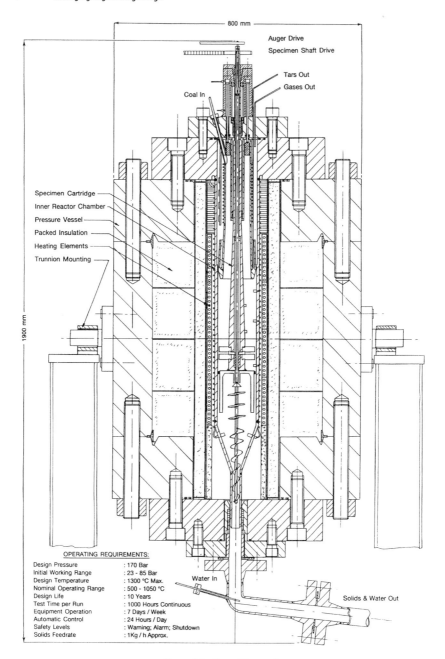

Figure 0.1. The gasifier test rig

defining the problem more closely and compiling a 20-page design specification or requirements list. This document, listing 308 requirements, formalized the input of everyone involved and recorded what had been agreed. Conceptual design, which was completed during the next 4 months, went smoothly. Embodiment design, involving the development of the reactor concept, the subsystem layouts, the control system design, and cost-justification documents, was completed during the following 17 months. Detail design of the seven subsystems, including the steelwork, was completed during the final 14 months. Six design reports were prepared during the course of the project, detailing all the technical design work involved. Field data recording the entire engineering design process for research purposes were collected from the time of the original proposal to the end of month 34, when the drawings were almost finished.

All references to the gasifier test-rig project have been taken from the doctoral thesis entitled *Analysis of the engineering design process in an industrial context*, submitted to Cambridge University by Crispin Hales in 1987 and first published by Gants Hill Publications in 1987, with a second edition in 1991.

0.2.2 The Life Chair

Repetitive strain injury (RSI) is a global work-related health problem. Disorders associated with RSI account for almost half of all occupational illnesses and, for example, have affected more than 5 million Americans to date. In 1991, a British Court awarded damages to two data-processing workers who suffered from RSI. Their employer, the British Telecom PLC, was ordered to pay UK£6600 to each of the plaintiffs. The company was found liable because it provided old chairs, which were difficult to adjust. The presiding Judge John Byrt concluded that "... the strain has been substantially added to by the strains which arose from the work systems in place and *poor posture due to poor ergonomics of the workstation, unadjustable chairs and....*" There is a need for the development of good ergonomic products for office work environments.

In 1956, a company was established in Petone, New Zealand, producing a range of products, including steel-framed furniture. In the late 1970s they designed and manufactured New Zealand's first office chair with a height adjustment. In the early 1980s the company changed it's name to Formway Furniture Ltd. In the mid 1990s, Formway identified an opportunity in the office furniture market for a more ergonomic chair. An eight-strong project team was established for the project, consisting of representatives from industrial design, engineering, marketing, management, and model making.

The Formway design team began by setting their task in context. Competitor products were evaluated and the influences on the project identified. This led to the development of a design brief in the form of a detailed list of project requirements. The crux of the problem was abstracted from the brief to be "*a support system for the human body, one that promotes movement rather than constraining it.*" Formway realized that the human body is not designed to sit, and that any posture, no matter how comfortable, loses its merit over time. Other key project goals included: environmentally sound design principles, compliance with standards and cost competitive/value for money.

Throughout the design, a systematic process was followed. The design task was divided into subsystems, and numerous concepts were evolved for each subsystem. Once the concept-generation process had been exhausted, the concepts were evaluated and final selections made based on the potential for meeting the requirements set out in the brief. The subsystem concepts were assembled to make concept variants for full chair assemblies. Ten full chair assemblies were prototyped and evaluated by the design team and by end users. "Blind" tests were conducted, where blindfolded office workers ranked the function and comfort of various prototypes and competitor products. This extensive testing program led to the selection of a final conceptual design solution for the chair.

The final solution was refined by detailed calculations to determine optimal structural properties, full consideration of available materials, and an in-depth study of manufacturing processes. This resulted in a final prototype, built using final materials and manufacturing processes (Figure 0.2).

Figure 0.2. The Life chair. *Courtesy of Formway Design*

Having developed a working prototype that satisfied the brief, Formway acquired patent protection for their novel ideas, while realizing that they did not have the manufacturing facilities or distribution capability to sell to a world market. For this, they would need the help of much larger company.

Knoll Inc. was established in New York in 1938 and is a worldwide leader in the design and manufacture of office furniture. Knoll has a network of dealerships, showrooms, independently owned dealers, and licensees in North America, Europe, Asia, and Latin America. They operate manufacturing sites in North America and Italy. Annually, Knoll sells about US$1 billion of furniture worldwide, with an 8.5% share of the North American market.

In October 2000, Formway signed a design license agreement with Knoll. The two companies embarked on a collaborative design effort to produce the manufacturing information allowing "Life" to be mass produced by Knoll. In June 2002, "Life" won a *Best of NeoCon Gold Award* for seating (desk/workstation/task chairs) at the prestigious NeoCon trade fair in Chicago.

The successful development of the Life chair can, in part, be attributed to the systematic design approach adopted, which exemplifies the approach advocated in this book. The "Life" chair project was selected as a contemporary case study for use as a common thread throughout the text.

PART 1
The Context

1 Ways of Thinking about Engineering Design
2 The Project Context

Chapter 1
Ways of Thinking about Engineering Design

1.1 Disasters and Failures
1.2 Engineering Excellence
1.3 New Innovations
1.4 Improving Engineering Design
1.5 Systematic Approaches to Engineering Design
1.6 Systematic Design in Practice
1.7 Tips for Management

1.1 Disasters and Failures

1.1.1 The Millennium Footbridge

Queen Elizabeth II officially inaugurated London's Millennium Footbridge on 9 May 2000. The bridge crosses the River Thames, connecting St Paul's Cathedral with the Tate Modern art gallery. Aesthetic design requirements called for a particularly low height profile, and the resulting design was the suspension bridge shown in Figure 1.1.

The bridge opened to the public on Saturday 10 June 2000. In the opening ceremony, more than 1000 pedestrians crossed the bridge, led by a band. This initiated a cyclic horizontal movement of the bridge deck with an amplitude of about 38 mm each way and a frequency of 1 Hz, which caused the walkers to stop as they tried to keep their balance. Following this harrowing event, pedestrian numbers were restricted, but the amplitude of the lateral movement was still considered a hazard to public safety. The bridge was closed on Monday 12 June, just 3 days after its public opening.

This raises the question: Should the design engineers have been able to foresee and prevent the Millennium Footbridge vibration? People-induced movement of bridges is not a new phenomenon. For example, in 1831, the 60th Rifle Corps were marching across the Broughton Suspension Bridge near Manchester, UK, when the bridge collapsed. As a result, the "break step" command was created as a safety procedure when marching soldiers across bridges.

A Millennium Footbridge research team was formed to solve what was commonly known as the "wobbly bridge" syndrome. The bridge's aesthetic "low height profile" requirement, together with its steel construction, resulted in a structure with low transverse natural frequencies and low inherent damping.

Figure 1.1. The London Millennium Footbridge

This combination of low frequencies and low damping made the bridge susceptible to people-induced movement. Similar experiences have been described in published literature, with explanations as to how medial/lateral body forces are transmitted to the ground as a person walks. In the case of the Millennium Footbridge, the frequency of this body force was close to the natural lateral bridge frequency. Results of the Millennium Footbridge study showed that, as the number of pedestrians increases, the group size reaches a critical value where the majority of the walkers begin to synchronize their sideways movement instinctively with the motion of the bridge. This synchronization promotes an increase in the vibration amplitude of the bridge structure. The "wobbly bridge" problem was overcome by fitting a combination of viscous dampers and tuned-mass vibration absorbers beneath the bridge deck. The bridge reopened to the public on 22 February 2002, almost 2 years after its aborted first opening.

Given the geometry and construction of the Millennium Footbridge, it would be surprising if a failure caused by people-induced vibration had not been considered by the designers. The costly and embarrassing problem may have been averted if lessons had been learned from the past. In the end, the detailed research for this particular situation has generated new information, which will allow better prediction of people-induced vibration for similar structures in the future.

You must learn from the mistakes of others. You can't possibly live long enough to make them all yourself.   Sam Levenson (1911-1980)

1.1.2   The Twin Towers Collapse

On 11 September 2001, two hijacked commercial aircraft were used to attack two of New York City's World Trade Center (WTC) buildings. One aircraft was flown into each of the two 110-story towers, resulting in the loss of 2830 lives. The subsequent collapse of both the WTC twin towers was undoubtedly one of the most costly engineering calamities of all time, and it raises questions such as:

- What caused the ultimate collapse?
- Would other skyscrapers with differently configured structures have collapsed?
- Could buildings be designed to prevent this type of catastrophic failure in the future?
- Should buildings of this type be designed to survive such an onslaught?
- How does this affect other current issues such as energy efficiency and global warming?

Each of the twin towers sustained substantial structural damage directly from the aircraft impact; yet, initially, the buildings remained standing due to the inherent robustness and redundancy of their steel framing system. This allowed most of the buildings' occupants to evacuate safely. A study commissioned by the US Federal Emergency Management Agency (FEMA, 2002) indicated that many buildings would have been more vulnerable to collapse and that, in the absence of other severe loading conditions, the towers could have remained standing.

Unfortunately, a second and more debilitating event occurred within each building, due to the fires caused by the aircraft impacts. Each aircraft carried a large quantity of fuel that ignited on impact, and much of this burnt off immediately in the form of fireballs exterior to the buildings. The FEMA report indicates that the amount of fuel that entered each building was not sufficient to produce the heat needed to collapse the structures. However, the fuel ignited the contents of the buildings, causing intense fires simultaneously on multiple floors. The immense heat input raised the temperature of the already compromised steel frames beyond the level at which failure would occur under normal loads, initiating the collapse of either the floor supports or the vertical columns in the impact zone (Bazant and Zhou, 2002).

The floors in the impact zone collapsed onto the undamaged floors below, under the weight of the undamaged superstructure above, as shown in Figure 1.2. The resulting impact load caused more undamaged columns to buckle, and the potential energy of the superstructure was rapidly converted to kinetic energy. A chain of progressive failures and a huge stress wave propagated down the structure, resulting in the total catastrophic collapse of each tower.

The collapse of the two WTC towers highlights the importance of designing buildings with sufficient strength and redundancy to provide extra strength capacity and alternative load transmission paths for survival in the event of significant building damage. It also highlights the dilemma faced by design engineers trying to produce a cost-effective design at an acceptable level of safety. Although the twin towers were designed with sufficient capability to accept the impact of a jet aircraft, whether intentional or not, the simultaneous ignition of office fires on multiple floors by burning fuel had not been perceived

Figure 1.2. The collapse of World Trade Center Tower 2. Courtesy of Thomas Nilsson/Getty Images

as a critical failure mode. As a consequence of this event, fire-protection studies are under way, and new resources for practicing fire-protection engineers may be evolved. It is likely that the jet aircraft load case will be considered in the design of buildings whose design or occupancy makes them susceptible to such incidents. New solutions, whereby buildings are designed to absorb impact energy, such as those proposed by Newland and Cebon (2002), if developed, could introduce another level of redundancy protection for buildings.

1.2 Engineering Excellence

It is an unfortunate fact that while design disasters are highlighted and remembered, excellent designs tend to be readily accepted into everyday usage and forgotten until something goes wrong. For example, consider the Crescent® Wrench, which was originally conceived in 1907 by Karl Peterson and marketed by the US Crescent Tool Company. It is a product that was so well designed originally that it has been used all over the world for decades with almost no change and has been copied by numerous manufacturers. Although it was developed long before the advent of computers and systematic design methods, it embodies a remarkable set of design features, which can hardly be bettered today.

A more recent example in the appliance industry is Fisher & Paykel in New Zealand, a company that has developed a series of technology platforms, such as their Smart Drive® clothes washers, Active Smart® refrigerators, and DishDrawer® dishwashers (Figure 1.3). These innovative products, which take full advantage of the microprocessor control of mechanical systems, are now gaining ground against more traditional products. The design excellence demanded by Fisher & Paykel has enabled the company to expand internationally from a strong Australian and New Zealand base.

1.3 New Innovations

Of course, the increasing speed of technological change opens the door to exciting new product innovations such as the Segway™ Human Transporter (HT), a two-wheeled, single-person transportation device invented by Dean Kamen. The Segway HT, shown in Figure 1.4, is self-balancing and features an intuitive control system where the user simply leans forward to move forward, and leans backward to move backward. To turn, the rider simply twists a turning grip on the handlebars to steer right and left (Heilemann, 2001).

It uses sophisticated control-system technology to maintain the balance of an inherently unstable device (an inverted pendulum). This is achieved using solid-state gyroscopes, which provide a reference for level balance much like the inner ear, and tilt sensors, which sense user input and acceleration. High-speed

Figure 1.3. The Fisher & Paykel DishDrawer®. Courtesy of Fisher & Paykel

microprocessors interpret the information and control electric motors, which provide a torque at the drive wheels that keeps the machine vertically stable.

The long-term commercial potential of the Segway HT remains to be seen, but the device has created immense public interest and curiosity. Although all new innovations are not always commercially successful, they inspire confidence in designers and encourage them to investigate and implement new technologies.

Figure 1.4. Dean Kamen's Segway®. Courtesy of Segway

1.4 Improving Engineering Design

Design is something that we all do one way or another, and we all think we could have designed things *better*. So we talk about it, analyze it, criticize it, argue about it and sue people over it. Much has been published on the subject, but it is dispersed within a variety of disciplines. Many design "methods" and "methodologies" have been developed, together with conceptual models, as techniques or aids for use in the activities of the design process, and more particularly for the *engineering* design process. Early developments were based on technical viewpoints that ignored many influences now regarded as important, and the approaches varied from country to country. For example, in the UK there was emphasis on innovative concepts, whereas in other European countries, especially Germany, it was more on quality through design. In North America the emphasis was more market orientated, with a systems or project management approach to design. Historically, engineering design has not been regarded as a major management concern. Managers tended to assume that *someone* would do "the drawings." The growth in international competition and the demand by users for better quality and safer products has forced changes in management thinking. In particular, this has been reinforced by the success of Japanese companies in raising and meeting customer expectations for high-quality products. Engineering design is now understood to be a key management issue, but not one that can be handled by bottom-line tactics. It is far more subtle than that.

1.5   Systematic Approaches to Engineering Design

In theory, the engineering design process is often described as a sequence of phases beginning with a perceived need and finishing with the detailed description of a particular technical system or product. Depending on the product, the phases may be labeled in different ways and carried out in parallel with the design of an appropriate manufacturing process. Each phase may be considered as a design process in itself, consisting of an iterative set of steps. Overall, and within each phase, however they are labeled, the engineering design process may be considered as a special case of "problem solving." Many design process "models" in the form of block diagrams have been developed to try and characterize the design process and so provide the design engineer with a somewhat defined procedure for applying available design techniques. However, until recently, such models and the associated techniques have been all but ignored in practice. Engineering design tended to be something that was just done by somebody in the *drawing office*.

During the 1970s there was renewed interest in the human activity of engineering design, and some more complicated *design activity models* were developed (Pugh, 1990). These represented a set of activities (such as marketing, task clarification, conceptual design, embodiment design, detail design, and manufacturing) within an overall management framework. In time, the use of terms such as *life-cycle engineering, concurrent engineering, simultaneous engineering* and *integrated product development* crept in, indicating a change of emphasis towards the parallel development and cross-linking of marketing, design, and manufacture. For the purposes of this book, the basic design process will be modeled as shown in Figure 1.5. It starts with an idea, need, proposal or brief,

Ways of Thinking about Engineering Design   19

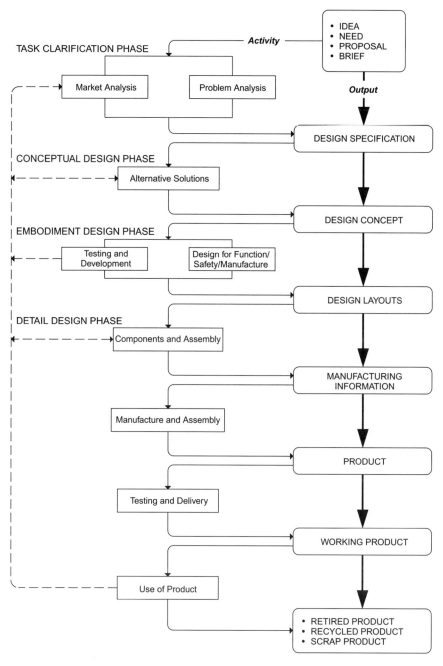

Figure 1.5. Schematic of basic design and manufacturing process

and progresses through the development, manufacture, and use of a product until its disposal, represented by the boxes on the right-hand side of the diagram. Each main phase in the life of the product may be regarded as the *output* from an *activity* or set of activities, represented by the boxes on the left-hand side and which vary depending on the type of product being designed. The broken lines on the far left represent necessary feedback loops between activities. Although the outputs are likely to follow each other in the sequence shown, the order and nature of the activities to achieve them may vary considerably from project to project. Iterations are common.

Note: it is assumed right from the start that the marketing, manufacturing, and quality assurance functions are represented within any engineering design team.

Two concepts that are of particular importance from the more holistic or *systems* approaches and which have now become accepted are: the concept of *resolution level* and the concept of *viewpoint*. Any system can be considered a subsystem within some larger system and, in turn, an umbrella system to its own subsystems. Whatever is defined to be "the system" at a particular time is a choice of resolution level or level of detail at which the activities are described at that particular time. At any particular level of resolution there are a number of viewpoints to consider. A corporate director may see a project in terms of the balance sheet for the company, whereas at another level of resolution the project manager might see it as a chance to develop a new product line. At the same level of resolution, but from a different viewpoint, the design manager for the project might see it as the development of an idea worth patenting. The concept of resolution level is illustrated in Figure 1.6. By mapping the context at different levels of resolution, it is possible to predict the likely influences on the project and to prepare in advance for coping with them. Defining, understanding, and working with the factors influencing the course of a project is critical to the successful management of engineering design.

1.6 Systematic Design in Practice

The justification for developing more systematic approaches to engineering design is principally that they will improve the quality and speed of design in practice. This development cannot occur in a vacuum. For them to be effective in practice, systematic approaches must also become accepted for teaching, training, researching, and the analysis of engineering design. They must become part of the engineering design infrastructure and be compatible with the use of computers throughout the design process. In practice, it is rarely possible, or necessary, to work through the whole design process on a particular project, yet the inherent features of a systematic approach can be applied to great advantage in almost every design situation. A systematic design approach provides a disciplined *way of thinking* that enables the design engineer to tackle any problem in a professional way, it provides a disciplined *way of working* that

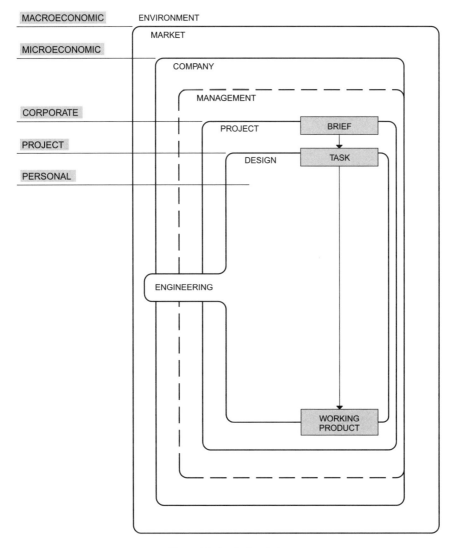

Figure 1.6. Levels of resolution

inspires confidence in management and in the customer, and it provides *working tools and techniques* to help ensure that quality solutions will be found within the constraints of the project. There is mounting evidence to show that if a systematic approach is *not* used, then the probability of a major disaster is high. For example, asking a simple set of questions, based on systematic design, would have alerted management to a developing disaster situation several *years* before the Space Shuttle Challenger disaster (Hales, 1987, 1989). The same holds true for many of the huge number of accidents involving design issues.

Systematic design does not imply a step-by-step serial process. Indeed, the beauty of a systematic approach is that it offers a framework for a complete project within which the design manager has room to move about, fitting together bits of the jigsaw as they come together and applying a variety of techniques to maintain steady overall progress towards a finished product. For example, component suppliers are often willing and able to provide design expertise with respect to their own specialties, and it pays to get them involved as early as possible, perhaps even before the conceptual design of much of the other equipment is finalized. Product literature on detailed components can often be helpful earlier in the design process than would be expected, sometimes even leading to ideas for broader concepts.

1.7 Tips for Management

- Treat engineering design as an important management issue.
- Learn from excellence in design, as well as from disasters.
- Consider new innovations that demonstrate the current state of the art.
- Try introducing a systematic approach to design as a framework for product development.
- Perceive the "big picture" through the concepts of resolution level and viewpoint.
- Develop the art of "windowing in" on tiny details and critically assessing their significance.
- Encourage disciplined ways of thinking and ways of working in design.
- Investigate what working tools are available and select the most appropriate.
- Be flexible in approach, take advantage of opportunities, and negotiate with persistence.

Chapter 2
The Project Context

2.1 Engineering Projects
2.2 Engineering Design in the Project Context
2.3 The Effect of Influences
2.4 Influences at the Macroeconomic Level
2.5 Influences at the Microeconomic Level
2.6 Influences at the Corporate Level
2.7 Design Context Checklist and Work Sheet
2.8 Tips for Management

2.1 Engineering Projects

Projects are a common denominator in engineering. During any engineering project, the design activities and the development of the designed system must be monitored. Every project is different, though certain types of project may have comparable features. What makes each project unique is the *context* in which it takes place. It is worth trying to map out the project context right at the start, and to be able to see the overall picture from different *viewpoints* and at different *levels of resolution*. Then we can choose particular levels of resolution, and look at the project from specific viewpoints. In this book we will be concentrating on the *engineering design* viewpoint.

2.2 Engineering Design in the Project Context

At the *project* level of resolution, typical phases of the work and the typical inputs and outputs may be represented as shown in Figure 2.1.

The project takes place within some kind of management system within an organization, generally a *company*. Typically, the company receives revenue from products being bought by customers in the *market*. A product is used by a user until its operational life is over. Customers and users are not necessarily the same, and often have different needs to be satisfied by the product. Once a product is established in the market, the revenue generated from it, less costs, provides the company with an operating profit until competition, demand, or new ideas makes it imperative for the expensive business of developing a new product by means of an engineering project. Naturally, it is in the interest of the

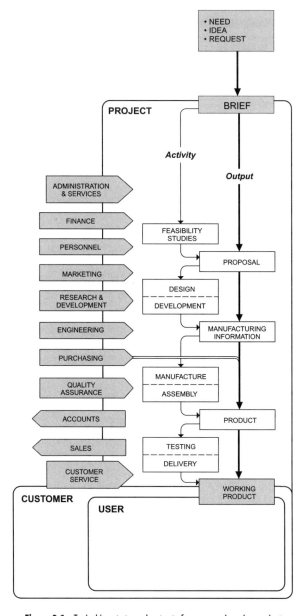

Figure 2.1. Typical inputs to and outputs from an engineering project

company to minimize the cost, time, and risk involved in such a project. For example, Japanese companies have developed what may be termed "incremental design," where new components or sub-assemblies are systematically introduced into an existing product to the point where there is almost a metamorphosis into an overall new and proven design. The "economic loop" within a particular market, as shown in Figure 2.2, may be used to identify and encourage the use of such approaches.

Each market exists within what might be termed an outer *environment*, which strongly influences what happens within the company, and hence what happens within a particular project. Figure 2.3 shows how we can now visualize a project, with its management, within a particular company, within a particular market, within the environment. Feeding into each project through individuals or groups are resources from the environment, the market, and the company. Customers, and thereby users, purchase products, generating revenue through exchange processes.

Within such a context we are concerned here with the *engineering* input to the project, as distinct from marketing, quality assurance, finance, or any of the other complementary inputs. By highlighting the engineering input, with both the design and production processes displayed as sub-sets within the project, the phases of the engineering design process may be visualized in terms of team activities and outputs, set in context with production, as part of a project within a company, within a market, within the external environment, as shown in Figure 2.4. This diagram is intended to function like the street map of a city. Although it may seem complicated at first glance, in just the same way as a street map it takes little time to become familiar with viewing it as a whole and then windowing in on the details as needs be.

The design process is often considered to be an iterative decision-making process. Although this is not a very accurate description of what actually occurs in practice, it is certain that without decisions there can be no progress through the steps and it emphasizes that management involvement (as a catalytic resource) is a crucial aspect of engineering design. Typical iterations in the process are represented in Figure 2.4 by the feedback loops. The transformation from "abstract ideas" to "concrete products" during the course of the design process is shown by changes in line-style around the loop as the information flow changes first to document flow then finally to material flow when manufacture starts. Thus, from the *engineering design viewpoint*, at this *personal* or design-team *personnel* level of resolution, the phases of the engineering design process may be simply described as follows:

1. Through *task clarification* activities the problem is defined.
   Output is a design specification.
2. Through *conceptual design* activities the solutions are generated, selected, and evaluated.
   Output is a concept.
3. Through *embodiment design* activities the concept is developed.
   Output is a final layout.
4. Through *detail design* activities every component is fixed in shape and form.
   Output is manufacturing information.

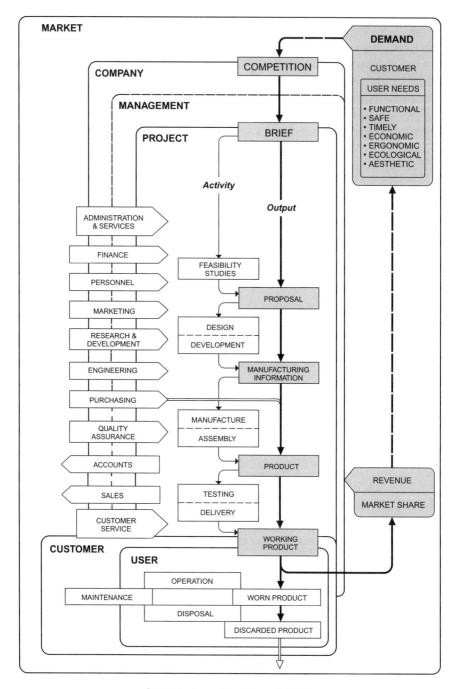

Figure 2.2. Economic loop for a typical project

The Project Context 27

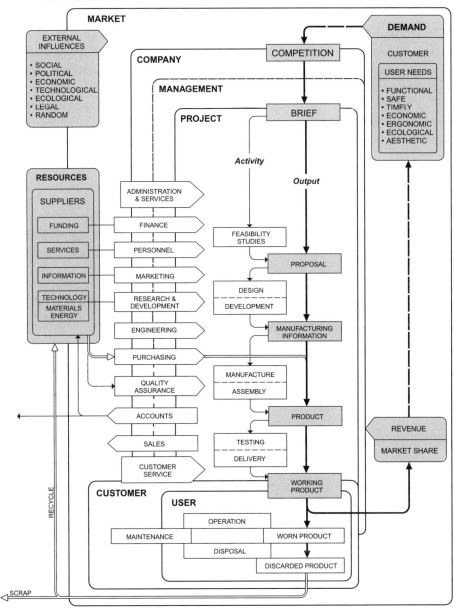

Figure 2.3. Project set in context

28 Managing Engineering Design

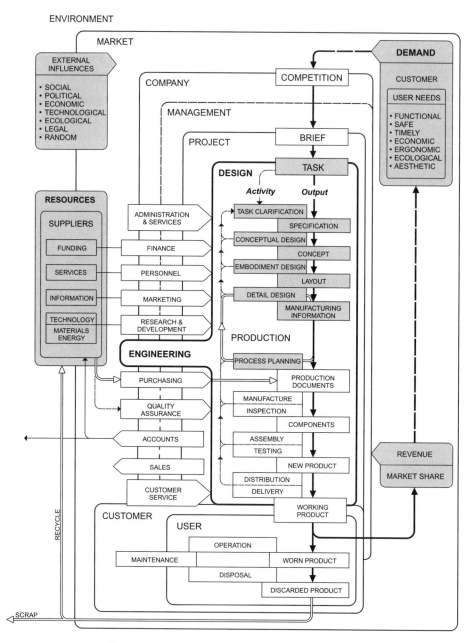

Figure 2.4. Engineering design process set in context within project

This conceptual model or map is useful in visualizing how the activities of design are influenced by numerous factors acting at different levels of resolution, and it is readily adapted to different project situations. For example, if a large company holds a monopoly in the market then the "company" may be regarded as equivalent to the "market." This is represented on the model by "windowing-out" the "Company" box to become coincident with the "Market" box while leaving everything else the same. The economic "loop" for the project then lies wholly within the overall company.

Having set engineering design in a general context, we can also window in on just the design process, as shown in Figure 2.5.

By "windowing in" and "windowing out" from one resolution level to the next, it is possible to concentrate effectively on the detail while keeping the wider context in mind, a crucial aspect of managing engineering design.

2.3 The Effect of Influences

One of the most frustrating things about being a design engineer or design manager is the way projects are manipulated by those who have very little to do with the design process itself. One minute everything is extremely urgent and the next minute the project is no longer required or the money has run out. More and more influences affect the course of design projects. It is necessary for the design manager to be aware of the impact of various influences and also to exercise some control over those that can be controlled while compensating for those that cannot, in the best interests of the customer, the project, and the design team.

Influences have been defined, for example by Lawrence and Lee (1984), as "people or things having power," with power as "the ability to affect outcomes." The engineering design process, as a goal-orientated process, cannot be effective unless the balance of power favors the attainment of project goals as distinct from goals at other resolution levels. A disgruntled project manager once said that his upper management had come out with an edict that project managers must become more "goal-orientated." And he had. His goal was to build up enough Frequent Flyer miles on company business to get a free round-the-world ticket!

Influences may range from being strongly positive towards the attainment of project goals, through neutral, to strongly negative. Also, they may be almost constant in effect, such as the pay scales for staff, or they may be highly variable, such as the degree of commitment to the project from an undecided management. At each resolution level there is a mixture of slowly changing "structure-orientated" influences, such as corporate organization, and continuously changing "process-orientated" influences, such as "enthusiasm" and "involvement." Though it may not be possible to define such influences as constants and variables in a quantitative way, it is certainly possible to identify

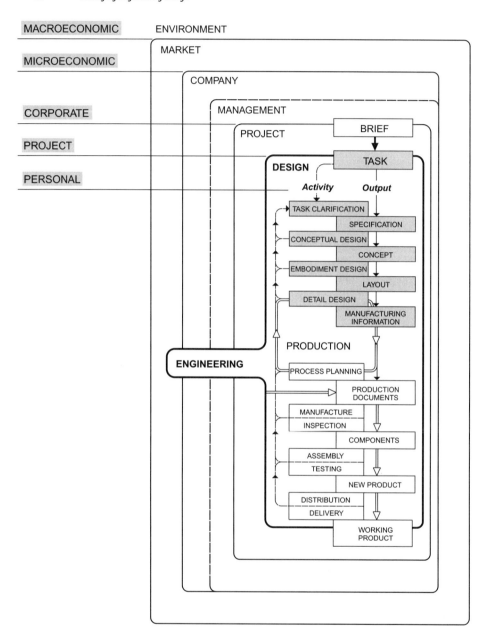

Figure 2.5. Levels of resolution related to engineering design process

categories of influence and contributing factors within each category, with subjective assessments of their observed impact.

During the 1970s, the Hughes Aircraft Company (1978) did a 5-year study in the USA on improving research and development productivity. This resulted in a practical set of checklists and guidelines for the compensatory control of influences. The engineering design process is analogous to the research and development process, and a study was carried out to identify a similar comprehensive set of influences specific to the engineering design process (Hales, 1987). The results of this design research have now been converted into a Web-based checklist and work sheet form to help the manager identify key influences in a particular design situation, then to monitor and deal with them in a somewhat systematic and controlled fashion.

2.4 Influences at the Macroeconomic Level

2.4.1 Cultural, Scientific, and Random

Design deals with the future and therefore is highly susceptible to cultural, scientific, and random influences. The factors contributing to these influences vary from country to country and from culture to culture in complex ways, as discussed in *The Seven Cultures of Capitalism* by Hampden-Turner and Trompenaars (1993). Important contributing factors within the category of cultural influence are social issues, the political climate, the economic situation, and legal requirements. These may be stable at a particular time and therefore have little effect on a project, but they can also change rapidly and leave the manager in an untenable position. For example, in 2001, the aftermath of the 11 September terrorist attacks in the USA had an immediate negative influence on many projects within the aviation industry. Social and political relations between American and Middle Eastern cultures were affected and the consequent slowdown in travel and tourism put the future of prominent airlines in question. Such influences are beyond the control of the project manager, but their effect may often be anticipated and compensating plans made accordingly. A manager of projects servicing the once stable aviation industry may now need to develop a contingency plan for alternative markets should the targeted market experience an unexpected downturn.

Scientific influences include the effects of technological developments and increasing concern with ecological effects in the environment. These are continually changing, and they will always have an important influence on the design process. Consider, for example, the effect of the first electronic wrist watches on the traditional Swiss watch-making industry, or the effect of recycling efforts on the design of aluminum cans for drinks. Technological developments tend to go in cycles and to follow "S-curves," as illustrated by the slide rule and electronic calculator example in Figure 2.6. The slide rule was developed to a highly sophisticated level by the end, and the very first electronic cal-

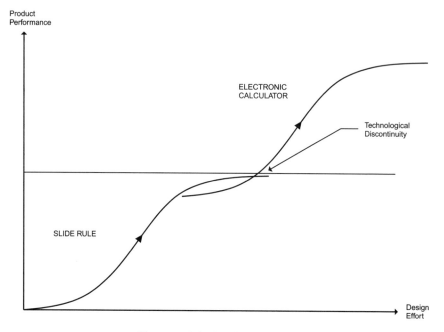

Figure 2.6. Technological cycles (S-curves)

culators were no match for the best slide rules. However, within a short time the capabilities of the calculator far surpassed that of the slide rule and the price of calculators steadily fell to the point where the slide rule became extinct. By the use of trend studies, expert opinions, and other means of technological forecasting, it is possible to predict some developments that might affect the project, but the design manager must always be on the lookout for that new idea which might wipe out the whole project at one blow. The notion of using "appropriate technology" (Schumacher, 1973) for the particular situation is also important. It is all very well developing a complex mechatronic water spray device using ultrasonic misters and oscillator circuits to keep vegetables fresh in supermarkets, but why do that when a system of simple valves and timers is just as effective? One such mechatronic system was so difficult to clean that it never was cleaned, with the result that bacteria in the mist caused an outbreak of Legionnaires' disease, followed by costly lawsuits. There is also the notion of "intermediate technology" (Intermediate Technology Development Group, 2003), where the level of technology used in a design is matched specifically to the capabilities and available resources of the users. An example would be the design of water turbines for small-capacity, low-head hydroelectric power-generation using only flat steel sheets (Giddens, 2003). Very little power-generating capacity is sacrificed by using flat blade geometry instead of curved blades, but there is an enormous advantage for poor communities attempting to develop self-sufficiency for the future, as the machines can be manufactured on site by local people, using local materials and simple tools.

Random influences are not controllable, but the effect of them on the project can be minimized by anticipation. They include the effect of "luck" and "chance," and a useful approach is to try and maximize the benefits of good luck while minimizing the effects of bad luck. For example, a design team may suddenly be offered the services of a highly skilled person, laid off from another project. It is not so easy to absorb additional people suddenly into a team, no matter how good the person is, but if the manager has thought about such a possibility ahead of time then advantage can be taken of such a situation. Similarly, if a key member of a design team becomes ill or leaves for some reason then it can devastate a project, but if contingency plans have been made ahead of time then the effect can be minimized. For example, it may be necessary to hire a replacement person under contract, and if such a person had been identified beforehand then it would shorten the disruption time.

Example: Gasifier Test Rig

At the start of this project, the political and economic forces in the UK favored development of coal gasification as an alternative energy source, and within the company there was emphasis on coal gasification research. This emphasis faded during the course of the project due to changes in Government policy for the purchase of natural gas from Europe, at prices making the use of synthetic natural gas (SNG) uneconomic well into the future. By comparison with these political and economic influences, the social, technological, ecological, and legal influences were insignificant. However, if construction of the rig had gone ahead as originally planned then the balance of external influences would have changed. For example, the immediate area around the company's property was being rapidly developed from a run-down industrial zone to an "up-market" residential zone, and there was increasing pressure on the company to ensure that it released no pollutants. The gasifier test rig would generate a small volume of hydrogen sulfide and, despite inclusion of an efficient gas scrubber in the system design, additional precautions for operation under emergency conditions were being discussed.

Random influences affected the project in many small ways. An example was the chance interchange between the contract design engineer and a company director for SNG production. Despite his lack of support for the gasifier test-rig project, the director said that he had passed the reactor assembly drawing on to one of his senior engineers who had commented favorably on a number of technical features. This gave some welcome encouragement in month 26, just as a final push on embodiment design was beginning. Bad luck also took its toll. The most significant event was the hospitalization of the contract design engineer due to peritonitis in month 16, just at the end of the conceptual design phase when the A-Form (cost justification) was to be submitted.

2.5 Influences at the Microeconomic Level

2.5.1 Market, Resource Availability, and Customers

The purpose of design is to address some kind of need, and unless it is clear what this need is, where it has come from, and the likelihood of it continuing as a need, the design manager runs the risk of designing something nobody wants or designing something to meet the wrong need. It is important to get as much information as possible about the market influences before the project starts, and also to monitor them closely during the course of the project in case things change.

Obviously, the outcome of a design project is bleak unless the market exists for the new product, or one can be generated through demonstration of the superiority of the product over existing equivalents. How to create successful products is an issue that has been analyzed by researchers such as Cagan and Vogel (2002), whose resulting simple approach for developing "breakthrough" new products has proven effective in practice. Only under exceptional circumstances, such as legislation mandating use of the product, will a mediocre design survive in the now highly competitive world markets. Even then it is likely to be rapidly superseded by improved designs developed by other companies. *Product planning* is, therefore, important: the systematic search for promising product ideas, together with their selection and development. Figure 2.7 is an attempt to show the project context from the *marketing* point of view, as distinct from the engineering design point of view. Marketing involves, for example, market analysis, discovery of new ideas, selection of appropriate product ideas, and the definition of particular products. It is essential that the design team draw in the expertise of the marketing staff right at the beginning of the project, to ensure that the technologically marvellous project will not turn out to be a financial disaster. Honest communication is absolutely essential. It is no use the marketing staff promising more from a product than is realistically possible, or the design team promising the product within unrealistic times or costs. There has to be a build up of mutual trust based on appreciation and understanding of the different points of view.

Resources are often a sore point between design teams and management. One reason for this can be seen in the familiar graph of a typical product life cycle, as shown in Figure 2.8. Design work is always a heavy cost item for a company, and it directly affects the cash flow in a negative way. It is quite possible to imagine the feelings of management towards a design team following the cash flow curve in Figure 2.8 as more and more money is spent with apparently little to show for it. This is most unfortunate, because without high-quality design a company is doomed, and cutting back on the resources of a design team is one sure way to achieve poor-quality design. Assuming that a project has been approved as viable, the design manager then has a major role to play in negotiating to get the best possible resources for the design team. By this is

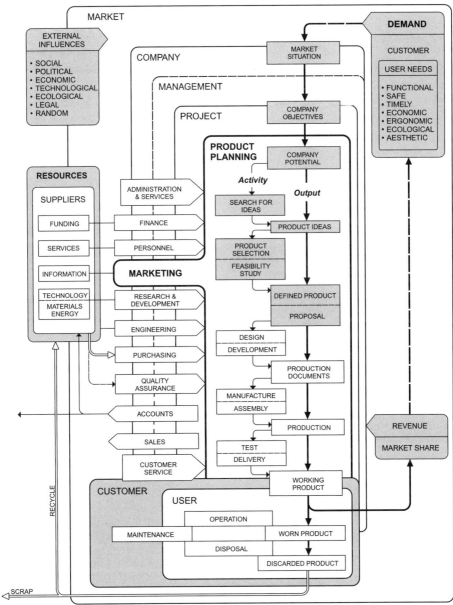

Figure 2.7. Marketing set in project context

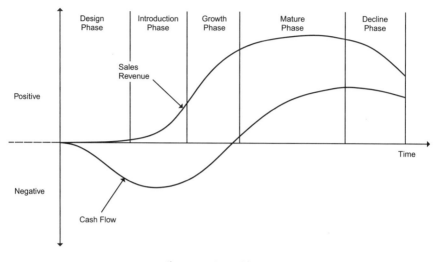

Figure 2.8. Project life cycle

meant the best possible people, the best possible funding, the best possible information, the best possible technology, the best possible working environment, and the best possible support all round. If, for *any* reason, the design manager fails to negotiate sufficient resources for the project then it will almost certainly cause the project to fail, though this may not become apparent until it is too late for recovery. The design team will lose heart and have difficulty maintaining respect for its leader. "Guesstimates" will be offered in place of calculations, sketches in place of drawings, sick time in place of overtime. Corporate management will demand results, requests for additional time will be met by extremely loud voices (America) or extremely quiet voices (Europe), and discussions over design issues will give way to recriminations over time and money. Most design managers have probably been through this sort of troublesome experience and come out of it wondering why they put so much effort into something that nobody seems to want in the end. Sooner or later one gets a feeling of frustration, and the thought of a becoming a sales manager seems rather attractive, with a company car and an air of breezy confidence at the positive end of the cash flow curve.

However, if the design manager instills a systematic approach into the design process then it will be found to have great advantages when it comes to the matter of resources. Corporate management will be more directly involved with the project, as the design process will be more visible, there will be tangible output to share and discuss during each phase, and the design manager will gain additional respect and goodwill from the professionalism demonstrated. Resource needs can be more precisely defined, problem areas identified sooner, and the whole design process managed in a less volatile manner.

> **Example: Gasifier Test Rig**
>
> Sufficient resources were available for the design effort, except for the lack of a qualified detail designer and a problem in obtaining field data on actual gasifier operating conditions. Unlike the control system design, where it was up to the project team to secure the services of a design engineer, detail design was under the control of a Services Group, and the recruiting of individuals for this was outside the control of the project team. When the time came for detail drawings to be done, no qualified person was available to do the work. What is more, it took a further 6 months to attract a suitable person and this caused a severe discontinuity in the project effort. The project had not been funded for construction, so the project team had little control over the situation. With regard to information needed on gasifier operating conditions, there was strict confidentiality on such information within the company. It was taken to such lengths that the rotational speed of a major component, essential for calculating the specimen movement in the rig, was wrong by a factor of 4 when told to the contract design engineer. The point here is not only that the contract design engineer wasted design effort because of wrong information, but also that this information was being used by permanent company staff in the absence of anything better.

2.5.2 Customers and Users

Ultimately, a product or technical system will be bought by someone and used by someone, and it is the perception of the value and the appeal to both customer and user that will largely determine the success or otherwise of the initial design. Therefore, it is an obvious first step to try and find out what the customer would like, but in practice this is not so easy. Customers often do not know what they want, and what they say they want is not always what they actually want. There is also often a difference between the real needs of customers and users. For example, carpet cleaning equipment may be purchased by a customer, Rent-a-Tool, but used by individuals who hire equipment from Rent-a-Tool. The rental company may look for particular features in the design, such as low cost or maintenance, in order to maximize the rental income, whereas the user is likely to be looking at it from quite another point of view, such as portability. The insurance company for Rent-a-Tool, on the other hand, is perhaps concerned about safety problems with the design from a liability point of view. Unless the design team is able to foresee how the design will be perceived by all the different parties involved in its use, and indeed what kind of *foreseeable misuse* it will be subjected to, there are likely to be many problems with the equipment during its life.

Another factor that influences design acceptability by customers and users is their continually changing expectations. Nowadays, pleasing the customer means aiming at a moving target. For example, sophisticated electronic calculators used to be something one would gather information about and have demonstrations on before getting close to a final selection. Now they are assumed to be bubble-pack items in a supermarket. Environmental and safety issues have become more and more important in design, and this trend is likely to continue, especially as the legal ramifications of noncompliance are enormous.

Example: Bicycle

On 30 January 1991, a woman in Wisconsin, USA, was paralyzed from a brain injury in a fall from a bicycle. Before trial her attorney negotiated a combined settlement of $7 million from the manufacturer, the importer, and the distributor. The lightweight, 15-speed bicycle was made in Taiwan, imported into America, and was given to the family as a gift when they bought a home entertainment center from the distributor company. The parents had given it to their only son, who then used it for the next 2 years. On the day of the accident the mother was riding this bicycle for the first time, as hers had a puncture. She fell off only a few blocks from her house, while going down a slight incline on a town street. People directly behind her and one person directly across the street witnessed what happened. Inspections of the bicycle after the accident revealed that the front forks of the bicycle were bent forward from their designed position, changing its steering characteristics. Although it was never established through evidence how the forks had become bent forward, and it was never proven that the bent forks were the cause of the accident, the design and manufacture of the bicycle was blamed for the whole affair. The manufacturer did not have sufficient records to prove that the bicycle met the agreed specifications. In addition, the importer and retailer were held to a higher standard of care than usual because they had copied a decal from another bicycle and put it on this one, in an attempt to promote it as competition caliber. The decal included the words: "CR-1010 Competition High-tension Steel." In engineering terms this means nothing, but it was held to convey a misleading message.

2.6 Influences at the Corporate Level

Although it is clear that the structure of a company and the way it works, or *organizational behavior,* has a great influence on engineering design projects, it is by no means clear what exactly the influencing factors are. Opinions and terminology in the management literature vary widely from author to author. Not only this, but the engineering design process is all but ignored in organization theories, though the production process is occasionally mentioned. Up to now the design manager has been left to develop effective management approaches alone.

By drawing from a wide variety of sources and gradually refining the huge number of influencing factors suggested in the literature, it has been possible to identify a condensed set of factors proven to be of importance with regard to engineering design. To make the list more manageable it can be broken down into small groups based, for example, on the "McKinsey 7-S Framework" described by Peters and Waterman (1982): corporate structure, systems, and strategy; shared values; and management style, skill, and staff. These groups form a sufficiently coherent set to be of use to us in assessing the effects of the organization on a particular design project. The condensed set of factors is discussed in the rest of this section.

2.6.1 Corporate Structure, Systems, and Strategy

There is a big difference between doing a design project within a large company spanning several countries and doing the same project in a small, perhaps family-run, firm. The large company is likely to have a wealth of resources in the way of facilities, specialists, and information. However, the access may be so cumbersome that the design engineer ignores it all and starts from scratch, getting information as if it has never been done before. It is common to find that an engineer on one side of the office does not know what others are doing on the other side of the same office. On the other hand, the small company is likely to be "lean" and close-knit but may lack the resources needed for a particular design project. The design manager in the large company might need to work towards better communication, whereas the design manager in the small company might need to find outside help to boost the necessary resources. By and large, in either type of organization, the more control the design team has over its own affairs the more likely it is to generate the enthusiasm, involvement, and tenacity to see the project through, but the team requires positive and continual encouragement from the upper management.

The way a company is organized may not be at all conducive to efficient or effective design, especially in the area of accounting. Design is such a wide-ranging activity that normal cost accounting systems and the thinking behind them often seem unable to cope. A simple example concerns telephones. Design engineers need to gather in a huge amount of information very quickly from all sorts of peculiar sources, and need to stay in close touch with many people. Effective design management would suggest assigning a direct-line telephone to each design engineer without restrictions or operator barriers. When cost accounting prevails it seems that the guard at the gate gets a direct dialling telephone for security purposes, but the design engineers have to plod through the operator from a shared telephone. The guard at the gate can phone day or night, but the design engineer has to call within operator hours. It is this kind of thinking that restricts design work and makes it almost impossible for a design team to compete internationally. A design engineer might need to be in the library one minute, calling Australia the next, making something out of a piece of wire the next, calculating something the next and negotiating for something the next. This is the essence of design, and anything put in its way is a barrier requiring extra time, energy, and money to get removed. Consider the thinking about books. A book may be cheaper than a good lunch and a lot more useful, yet the buying of a book often requires special management approval. Another example is the way time is accounted for. From the design manager's point of view the one tangible thing that can be measured is actual hours of work, and it then does not matter whether it is done in the middle of the night or at work on Monday. However, if the accounting system is based on days of work in an average week with certain average hours per day and a short day on Friday then you can either forget about doing design or forget about effective cost control. Design cannot be done in average days.

Example: Gasifier Test Rig

The project manager's monthly cost sheets were in terms of people rather than projects, and in terms of 1/10th days rather than hours. The measurement of project effort in 1/10th days would have been virtually impossible from a design research viewpoint, especially with Fridays having shorter hours than other days. Although an attempt was made to flag all the costs and effort attributed to the gasifier test rig by means of an extra digit on the job number, this digit was not recognized by the computerized accounting system. The project manager was surprised at the small number of total hours (2368) recorded by the participant observer: "It had seemed to be more than that," but an approximate check through the manager's cost sheets confirmed that the total project effort was about 1.5 "man-years".

Corporate management tends to consider pay scales and employee benefits as a "package," and perhaps this is the best approach for most employees. However, there are complications with regard to design. It may be true to say that the higher the pay scale the more motivated the design team is likely to be, but the matter of benefits is a problem. For example, flexi-time may be fine for certain types of employment but it needs to be carefully thought out with regard to design. If half the design team comes in early and leaves early while the other half does the reverse, then it is soon found almost impossible to get the whole team together at one time for some solid work output. Of course, the idea of giving design engineers more freedom is excellent, but unless those design engineers have the project as their first priority then this personal benefit is very much to the detriment of the project. Similarly with holidays: we would all like the luxury of long vacations, but unless there is some control over when they can be taken then the project can suffer greatly.

Example: Gasifier Test Rig

Both the pay and the benefits offered by the company were considered good by most team members, and in the case of one or two they were the main reasons for them staying in their jobs. From the gasifier test-rig viewpoint, however, the influence of pay was quite different from the influence of benefits. Whereas the level of pay was observed to act as an incentive, particularly with the contract staff, the benefits in the form of vacation time, holidays, "sick time," "flexi-time," and personal freedom were observed to cause unpredictable disruptions in project progress. The type of problem this caused within the project team is illustrated by a notebook entry on 9 April: "Holiday schedule: J__ in until 19th, then away 1 or 2 weeks; R__ in until Easter; F__ away 16–27 April and again 13 May to 23 June; H__ away 2 weeks after next week; Easter Holiday 20–23 April; Bank Holidays 7 & 28 May."

2.6.2 Shared Values

In a sense, all this comes back to the attitude and approach of the corporate management. If the management make their objectives clear, make it clear what risks are being taken, make it clear that they are committed to the project, and transfuse their enthusiasm through active involvement, then the design engineers are likely to respond in a positive fashion and not take personal advantage of benefits to the detriment of the project. The design manager is caught in between, and must see things from both points of view so as to motivate everyone in the direction most beneficial to the project. This is easier said than done in an economic climate where trust in management has gone, and loyalty is history. How is a design manager supposed to remain enthusiastic at a project level when the corporate executives may, at any time, uproot not only their traditional manufacturing facilities, but also their design capability, and move it all from country to country in their continual quest for cost reduction?

2.6.3 Management Style, Skills, and Staff

In a simplistic way, we can look at the extremes of management style as follows:

- Autocratic – what the boss says goes.
- Benevolent – what the team says goes.
- Consultative – what the boss says goes, after others have been heard.
- Participative – what the boss and team say together goes.

There are advantages and disadvantages with each of these extremes, and it depends on what type of project is being carried out as to which style, or mixture of styles, is likely to be the most appropriate. It took an autocratic style to produce the Sony Walkman; it took a combined participative and consultative style to produce the Life chair. For any particular project, a design manager has to assess whether the degree of design-team freedom and the degree of design-team participation is appropriate, and what to do about it if it is not.

> **Example: Gasifier Test Rig**
>
> Of the four styles (autocratic, benevolent, consultative, and participative), the benevolent style was most in evidence. It was observed at all levels of management. Concern for an employee's personal problems and health sometimes took precedence over concern for the project, and personal vacations could be scheduled at any time. "Flexi-time" gave additional personal freedom, and the working atmosphere was generally relaxed. Thus, the predominantly "benevolent" style of management tended to favor the team members at the expense of the project, and this acted as a negative influence as far as project progress was concerned.

The design team has a tough job to do and it needs the support of quality management, *i.e.* management with the skills to ensure that things are planned out, coordinated properly, and with adequate resources available at the right time. There has to be keen interest in the project and an element of the "project champion" present to boost confidence in the project on behalf of the project team. A design team is expected to be *effective* (doing the right things) and *efficient* (doing things right), but to accomplish this the team needs managers who can communicate well, who have good judgment, who are motivated themselves, and who have sufficient confidence in themselves to guide the team all the way from design specification to working product.

Example: Gasifier Test Rig

The 5-month period of indecision regarding funding of the project would suggest that, at the time, the corporate strategy on coal gasification research was not clear, at least not to those responsible for approving funding for the gasifier test rig. It also indicated a reluctance to take risks. To proceed with the detail design work but not the application for construction was a way of "hedging one's bets." These were important factors, as a slightly clearer strategy might have forced the decision against the project much earlier, and a slightly less cautious approach certainly would have favored construction. In the literature, "innovation" (implementation of a design or new ideas) is seen as an important influencing factor at the corporate level. The gasifier test rig was regarded as "novel" in design, but until it was built and operating it could not demonstrate "innovation"; so, although this contributing factor was considered important, the project data could provide no evidence for this. It would seem that innovation and risk taking are interdependent: had the more risky decision to build the rig been taken, and had the rig performed as expected, then it is likely that the project would have been seen as innovative. Another factor often stressed in the literature is corporate "involvement." For this project, such corporate involvement (*i.e.* higher level than project management) was intermittent, and it was either at the request of the project team or as a result of a chance interchange. No unsolicited corporate involvement was observed; and, as far as the project team was concerned, this was seen to indicate a lack of commitment towards the project, acting as a negative influence.

2.7 Design Context Checklist and Work Sheet

To assist the design manager in building up a picture of the context within which a design project will, or is, taking place, the *Design Context Checklist* shown in Figure 2.9 has been developed, together with the associated *Design Context Work Sheet* shown in Figure 2.10. These are provided as electronic files

DESIGN CONTEXT CHECKLIST

LEVEL	INFLUENCES	CONTRIBUTING FACTORS	SOME QUESTIONS TO ASK: EFFECTS ON PROJECT?
MACRO-ECONOMIC	CULTURAL	Social issues Political climate Economic situation	Effects of social change? Effect of politics? Effect of economic situation?
	SCIENTIFIC	Legal requirements Technological advances Ecological concerns	Regulations, codes, standards, liability? Changes in technology? Environmental problems?
	RANDOM	Luck/chance	Effect of luck/chance?
MICRO-ECONOMIC	MARKET	Demand Competition Financial risk	Demand for product? Competition for product? Effects of success or failure?
	RESOURCE AVAILABILITY	Human services Capital finance Information for design Appropriate technology Appropriate materials Appropriate energy	Right people available? Enough money for job? Enough design information? Do we have the technology? Access to materials? Power/fuel supplies adequate?
	CUSTOMER	Understanding of need Urgency of need Expectations Involvement	Is it clear what customer needs? Is there time to do the job? Expectations realistic? Customer helpful in design?
CORPORATE	CORPORATE STRUCTURE	Span of company Size of company Type of project control	Effect of company span on project? Effect of company size on project? Adequate project independence?
	CORPORATE SYSTEMS	Help getting information Quality of work environment Pay scales and benefits	Information easily obtained? Work environment good? Effect of these on project?
	CORPORATE STRATEGY	Clarity of objectives Level of risk taking/innovation	Does company know what it is doing? Is management strong/innovative?
	SHARED VALUES	Degree of commitment Degree of involvement Degree of project enthusiasm	Management commitment adequate? Management involvement adequate? Management enthusiasm adequate?
	MANAGEMENT STYLE	Degree of staff freedom Degree of staff participation	Is staff encouraged to be creative? Is staff involved in management?
	MANAGEMENT SKILL	Quality of planning/coordination Quality of communication Effectiveness of project support Effectiveness of resource use	Are the management plans realistic? Is communication effective? Is there a "project champion"? Are resources used effectively?
	MANAGEMENT STAFF	Number of staff involved Quality of judgment Degree of motivation/morale Degree of confidence	Is there enough input from staff? Is good judgment exercised? Sufficient motivation/morale? Is confidence high?

Figure 2.9. Design context checklist

The Project Context   45

DESIGN CONTEXT WORK SHEET

PROJECT: _____   DATE: _____

LEVEL	INFLUENCES	CONTRIBUTING FACTORS	CURRENT STATUS (Positive / Neutral / Negative)	REQUIRED ACTION (Compensate / Promote / Disregard)
MACRO-ECONOMIC	CULTURAL	Social issues / Political climate	☐☐☐☐☐	☐☐☐
	SCIENTIFIC	Economic situation / Legal requirements / Technological advances	☐☐☐☐☐	☐☐☐
	RANDOM	Ecological concerns / Luck/chance	☐☐☐☐☐	☐☐☐
MICRO-ECONOMIC	MARKET	Demand / Competition / Financial risk	☐☐☐☐☐	☐☐☐
	RESOURCE AVAILABILITY	Human services / Capital finance / Information for design / Appropriate technology / Appropriate materials / Appropriate energy	☐☐☐☐☐	☐☐☐
	CUSTOMER	Understanding of need / Urgency of need / Expectations / Involvement	☐☐☐☐☐	☐☐☐
CORPORATE	CORPORATE STRUCTURE	Span of company / Size of company / Type of project control	☐☐☐☐☐	☐☐☐
	CORPORATE SYSTEMS	Help getting information / Quality of work environment / Pay scales and benefits	☐☐☐☐☐	☐☐☐
	CORPORATE STRATEGY	Clarity of objectives / Level of risk taking/innovation	☐☐☐☐☐	☐☐☐
	SHARED VALUES	Degree of commitment / Degree of involvement / Degree of project enthusiasm	☐☐☐☐☐	☐☐☐
	MANAGEMENT STYLE	Degree of staff freedom / Degree of staff participation	☐☐☐☐☐	☐☐☐
	MANAGEMENT SKILL	Quality of planning/coordination / Quality of communication / Effectiveness of project support / Effectiveness of resource use	☐☐☐☐☐	☐☐☐
	MANAGEMENT STAFF	Number of staff involved / Quality of judgment / Degree of motivation/morale / Degree of confidence	☐☐☐☐☐	☐☐☐

Figure 2.10. Design context work sheet

on the CD accompanying the book, for Web-based use within geographically dispersed design teams. The checklist provides a list of questions to ask oneself, the design team, upper management, or others, and the work sheet provides a series of answer boxes to fill out. Both the checklist and work sheet are broken down by level of resolution, area of influence, and contributing factor. The work sheet has an assessment column for recording whether the influence factor is considered negative or positive with regard to the project, and how strongly. Then there is an action item column for the design manager to decide whether to try to control or manipulate the influence, compensate for it, or simply monitor it and hope for the best.

The completed work sheet becomes a status report on the key influences impinging on the project at that time, as shown by the example in Figure 2.11 based on the reconstruction of the Formway Life chair project. Influences at the *macroeconomic level* shaped the design intent for the Life chair. With reference to Figure 2.11, contributing factors were as follows:

1. *Social issues* and *technological advances* were closely linked and positively influenced the chair project. The team predicted areas of likely social change due to technological advances in office working environments and the need to accommodate users working "away from the office."
2. *Legal requirements* were considered to have a positive influence on the project. The majority of competitor products did not meet the ergonomic needs of the end user; hence, when public perception of the health and safety risks reaches a sufficient level of intensity, this is likely to be reflected in legislation. A change in health and safety legislation would favor products with superior ergonomics. Legal requirements were made a high priority ("promote" on work sheet) and ergonomics became a primary driver in the design of the chair.
3. The *political climate* could be either strongly positive or strongly negative and was considered unpredictable by the design team. The status needed constant monitoring and was compensated for by obtaining resources from suppliers in different political climates.
4. The *economic situation* was positive because the weak New Zealand dollar at the time generally favored export goods.
5. *Ecological concerns* were seen as having a positive influence on the project. A focus on environmentally sound principles builds on New Zealand's "clean green" image and, if promoted, would give the product a competitive advantage from both a customer and legal perspective.
6. The effect of *luck and chance*, due to the detailed brief and meticulous project planning, was considered as neutral by the design team. They did, however, promote this effect, *e.g.* by taking advantage of opportunities to use new technology.

Influences on the project at the *microeconomic level* led to the development of the primary market goals for the Formway Life chair project. Significant influences (Figure 2.11) were as follows:

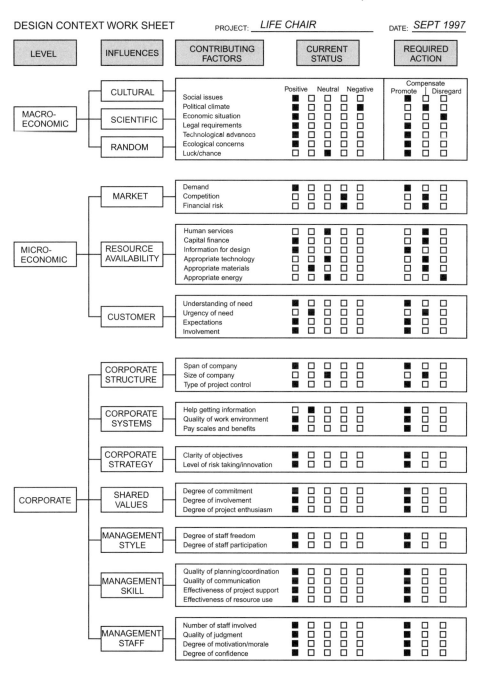

Figure 2.11. Example of design context work sheet

1. The *market influence* was considered positive because the demand for a cost-competitive ergonomic chair was identified as a likely continuing market requirement. Market research identified a promising opportunity for a chair that was both value for money and had superior ergonomics.
2. There was strong *competition* in the market. Despite the positive market influence, the influence of competition was negative because, as the Life chair was being developed, competitors were also evolving better products and patenting ideas that impinged on Life's competitive advantage in the market.
3. The *financial risk* of developing the chair was seen to have a negative impact on the market influence. Formway's manufacturing facility was too small to manufacture and distribute this product internationally. This factor was compensated for by the smaller New Zealand company (Formway Design Studio) licensing their design at the working prototype stage to a much larger American company (Knoll Inc.) for the detail design, manufacture, and distribution phases.
4. *Human services*, *appropriate technology*, and *access to materials* were considered, overall, to have a neutral influence. On the one hand, there were positive influences, such as: the design team were very experienced furniture manufacturers; they had up-to-date appropriate technology; they had access to and experience with materials for manufacturing office furniture. On the other hand, there were negative influences due to the effect of novelty. Compensation for these factors was made by contracting external help, such as staff training, engineering analysis (*e.g.* three-dimensional (3D) scanning), specialist engineering machinery (*e.g.* rapid prototyping), and advice on the use of new materials.
5. *Capital finance* was strongly in the design team's favor during the task clarification, conceptual design, and embodiment design phases. A realistic budget was set aside for the design phase; however, the detailed design phase capital finance was compensated for by the collaborative partnership with Knoll Inc.
6. *Information for design* was considered a positive influence. The team members were able to use whatever means possible to gather design information, *e.g.* team members had the opportunity to visit international trade fairs.
7. The *customer* was considered to have a positive influence on the project. The customer's *needs* were clear and their *involvement* in user trials was likely to be effective in prototype assessment.
8. The customer's *expectations* were high; however, this had a positive influence because it was perceived that these expectations could be easily met.
9. *Urgency of the need* was considered neutral. The design team were given sufficient time to complete the task within the company; however, this was offset by the constant threat of a competitor in the market introducing a new product or patenting new ideas first.

Influences at the *corporate level* provided critical guidance and support for the design team. Contributing factors (Figure 2.11) were as follows:

1. The *corporate structure* had a significant influence on team dynamics. Formway Design Studio is a small company with around 15 designers. The design team operates out of a single office alongside two other small project teams. In this case, the *span of the company* resulted in a close-knit group, which positively influenced communication between team members. On the other hand, the *size of the company* had a negative influence in terms of obtaining resources; this was compensated for by building collaborative relationships with companies who had specialist skills and specialist equipment.
2. The *corporate structure* allowed the design team adequate independence, and hence control, over their project working. This *type of project control* promoted a level of freedom that encouraged initiative in daily project working.
3. *Corporate systems* allowed designers unrestricted access to all available communication tools and, when necessary, *help in getting information* was compensated for by employing external consultants.
4. The design team worked in an open office with team members grouped according to specific design activities; the office was fitted with high-quality furniture throughout and there were areas set aside for social interaction. This promoted a fun *work environment* with good communication (and healthy banter) within the team, but there were also quiet areas where individuals and small groups could work uninterrupted. A communal project work area was established for this particular project so that team members could exchange ideas using a white board, post design information on a notice board, or hold project meetings.
5. *Pay scales and benefits* were considered to be at a good level, and designers were permitted to work flexible hours; however, time was scheduled where all team members or specific groups were required to be available in the design office at one time.
6. The *corporate strategy* was transparent to the design team; hence, management's *objectives* were *clear*, and this had a positive influence on the project.
7. The company's management had considerable experience in supporting the development of office furniture, and this supported the high *level of risk and innovation* required to evolve a new chair concept.
8. There was a strong sense of *shared values* between management and the design team. The *degree of commitment* by management in providing the necessary resources strongly favored this project. In fact, this was considered by the team to be almost to the detriment of other internal projects.
9. Representatives from *management were involved* in all critical project decisions, and there was considered to be a high level of *project enthusiasm* by management; these factors had a strong positive influence on the design team.

10. *Management style* was predominantly *participative*, where representatives from management and the team members had an equal say in deciding the direction of the project. This kept the project on track while generating a sense of ownership that led to an enthusiastic design team with the tenacity required to accomplish project milestones.
11. Team members were allowed 1 day per week of absolute *freedom* to pursue their own project ideas. This encouraged creativity and, although team members were not required to direct this free time towards any particular project, their enthusiasm for the Life chair concept was evident, in that they generally elected to spend their free time pursuing ideas for the chair. The team was also encouraged to "get out of the office" and try new environments for stimulating ideas. For example, the team would often pack up a portable white board, some good food, and drinks and go to a local boat club for brainstorming sessions.
12. Management's *planning/coordination* was detailed and was perceived as realistic by the design team. When goals were not found to be realistic, the management then showed understanding in its approach to the negotiations for revised realistic goals.
13. *Communication* between management and the design team was effective due to: the availability of management (on site with an open-door policy); a management team that was considered approachable by design team members; management being proactive in informing and involving the team in decisions that had a bearing on their project working.
14. The *effectiveness of project support* had a positive influence due to two of the company directors being considered as part of the design team and "project champions" at the management level.
15. *Resources were used effectively* by management to progress the project, and this had a positive effect on the project.
16. The *number of staff involved* was a positive factor. Together with the design team, representatives from marketing, finance, and production were all considered as stakeholders in signing off at project milestones. The inclusion of all the stakeholders ensured that good judgment was exercised in guiding the project to obtain successful outcomes.
17. The *degree of motivation and morale* had a strongly positive influence on the project. This was due to the positive attitude of management, commending the team for good work and showing their appreciation by celebrating achievements.
18. Management understood the strengths and weaknesses of the design team and, hence, was able to demonstrate a high *level of confidence* in the project team.

In looking at the context work sheet, it is not hard to see why this project was successful. The team used the macroeconomic influences in a positive way; this promotes a final product that is likely to be stable in its intended market. At the microeconomic level, the market opportunity was defined, the resources availability established, and it was shown that the needs of the customer could be met. At the corporate level, the design team and the management team worked together to achieve a common overall objective.

The completed work sheet can be referred to in progress or review meetings, used for discussion purposes, and updated at regular intervals. It provides the design manager with a simple way of keeping some measure of control over difficult issues in a systematic fashion and recording what the thinking was at the time. The Web-based format provides easy future reference and helps in compiling company design histories.

2.8 Tips for Management

- Design projects take place within a specific context.
- Mapping the context helps in visualizing the "big picture."
- Keep the big picture in mind, then "window" in and out on details.
- Five levels of resolution provide a useful framework for the context:

    Macroeconomic level – environment external to the market.
    Microeconomic level – market within which the company is operating.
    Corporate level – company within which the project takes place.
    Project level – project with engineering design input.
    Personal level – individual/team inputs to design process.

- Identify and understand the different viewpoints at different levels of resolution.
- Design projects involve team activities, team outputs, and contextual influences.
- Activities and outputs of the engineering design process may be phased as follows:

    Task clarification activities result in a design specification.
    Conceptual design activities result in a design concept.
    Embodiment design activities result in a design layout.
    Detail design activities result in manufacturing information.

- Use the checklist to help identify key influences.
- Explore the influences on the engineering design process at each level of resolution.
- Summarize the positive and negative aspects on the work sheet.
- Take appropriate action and review on a regular basis.

PART 2
Task, Team and Tools

3 Profiling the Project
4 Managing the Design Team

Chapter 3
Profiling the Project

3.1 Influences at the Project Level
3.2 Design Task
3.3 Design Team
3.4 Design Tools and Techniques
3.5 Design Team Output
3.6 Project Profile Checklist and Work Sheet
3.7 Tips for Management

3.1 Influences at the Project Level

The design team and the way it works are critical to the outcome of any design project. A well-chosen and carefully managed team is essential, but it is often difficult to create an effective team from the available resources. Design projects have widely varying characteristics, and it is necessary to create a team responsive to the particular set of characteristics. In any particular project, the team composition may need to be modified as the project progresses through its natural phases, but the manager may not have the power or the resources to make the necessary adjustments. The best that can be done is to assess what sort of team would be ideal for the project and match this as closely as possible with the available people and available services.

A first step in this matching process is to consider the characteristics of the project in more detail. One way to do this is to take the four main features of a project as described by Rodwell (1971) and build on these to compile a comprehensive *project profile* covering the task, the design team, the design tools and techniques available, and the design team output. This may then be used to monitor progress and assist in matching the design team characteristics to the task as the project progresses through its different phases.

3.2 Design Task

The four main characterizing features of a project that Rodwell (1971) identified were:

- Magnitude
- Complexity

- Novelty
- Production quantity.

Project *size* is an obvious factor, but it has some subtle effects. On a small project, the project plan, and hence the path that the project follows, may be clearly defined and more transparent to the design team. Each of the team members may be involved in a wide range of tasks and they may know the intimate design details of every component.

In a larger project, a higher level of coordination and project planning is required. The "big picture" may not be as clear to each of the team members. Individual team members are likely to be working on single components or subassemblies of a much larger system. The design tasks are often more specialized and there are likely to be more resources available to complete tasks to a much greater level of detail. For example, one of us was once involved in a project, which from his viewpoint was a large project, requiring a much larger team than for previous projects. An outside consultant working as part of that design team had exactly the opposite problem. All his life he had worked on large-scale projects, such as blast-furnace design, and from his viewpoint this was a tiny project. So here was a team where one person was trying to come to terms with working on the largest project he had managed before, while another was trying to come to terms guiding the smallest project he had worked on before. The result was not a happy one, as the consultant, who was highly respected in his field, was keen for certain features to be incorporated in the design, which, it turned out later, were on too "grand" a scale within the scope of the project. The author, on the other hand, did not have the experience at the time to assess the full impact of what was being agreed to and had too little "clout" to do much about it anyway.

Projects range from very *simple* to extremely *complex*, and it is important to get a feel for where the project lies along this scale. For example, the project mentioned above required a pressurized natural-gas feed system, an oxygen feed system, an excess air system, a slag feeding system, and a water cooling system, all on a large industrial scale but squeezed into a confined space within existing building facilities. This made for a highly complex design problem. Had more space been available the problem would have been greatly simplified and the demands made on the design team would have been considerably reduced.

The level of *novelty* involved in a project also has a far-reaching influence on the way the team is built up. In many projects the level of novelty is low and it is possible to prove the technology before there is any manufacture. In this case there is little advantage in developing a team strong in creativity. It may be better to have a team with exceptional strength in detail design instead. For example, in vehicle design, many components stay the same from one model to the next and the overall concept rarely changes, whereas the design of equipment to do specialized testing usually involves some kind of completely new concept. The design team for the new car model needs to include staff from many different departments to ensure the integration of design and manufacture on a volume production basis, whereas with the design of specialized test equipment the important question is often just "will it work?"

Similarly, the *production quantity* has an influence on the makeup of the design team. If the product is to be mass-produced then clearly the economics of how it is to be manufactured are of critical importance, and the manufacturing process selected may dictate how the product is designed. However, with a one-off product or system the more important issues are usually whether or not it can be made at all, and if so, will it work to the customer's expectations.

Two additional influences on the design task are *technical risk* and *urgency* or delivery time. All design projects carry an element of technical, as well as financial, risk, but the magnitude of each varies widely depending on the circumstances. It is important that the design team works within the limits of technical and financial risk that it feels comfortable with, so as to avoid any unnecessary frustrations and aggravations that might detract from the team's productivity. The design manager has the difficult task of assessing whether or not the design team is capable of meeting the customer's expectations within the bounds of what is technically possible in the allotted timescale. If the design calls for the use of any unproven technology, or large scaling factors, then this should be a warning of problems ahead. The project may require more money, more time, or both. In a sense, all design projects are "urgent," as the finished product is wanted as quickly as possible, but often the timescale set for the design team is completely unrealistic and compromises must be agreed upon.

Example: Tri-axis Transfer Press

Two huge tri-axis transfer presses for making car body parts failed to perform adequately in service. A team of Triodyne engineers was set up to investigate the failures and analyze the design of the machines. Some interesting characterizing features of the project emerged:

- *Largest* combination line machines ever built up to that time.
- *More complex* transfer mechanism than on previous machines.
- *Novel* performance requirements specified, such as high-speed operation.
- *One-off*, custom-built machines.
- *High technical risk*, with complete separation of design from manufacture.
- *High degree of urgency*, with penalty clauses for late delivery.

The contracts called for the machines to be manufactured by the supplier company in Europe, but to be designed by a subcontracting company in

Continued

> North America. This meant that, in addition to the above features, some 4000 drawings had to be translated from one language to another, imperial units had to be converted to metric, and material specifications had to be matched according to different standards. The timescale was so short that the contracts permitted use of drawings from previous machine designs where practicable, and to reduce late-delivery penalties the machines had to be built directly in the customer's plant without prior testing. They were then put into production without any systematic commissioning or "shakedown."
>
> The failure of the machines to perform to the customer's expectations caused such loss of revenue to the customer that the claim for damages against the supplier company and the subcontracting company far exceeded the cost of design and manufacture of the machines. The supplier and subcontracting companies were both long-established and well-respected designers and manufacturers of large metalworking presses with a history of working together on previous contracts. Both companies are now out of business.

3.3   Design Team

Team building forms an important part of the design manager's responsibilities. People have a *functional role* in the team, using their particular technical expertise and experience, and obviously this has to be matched to the work in hand. They also have a *team role*, in the sense of using their particular character traits to help make the team as a whole work as a team. If the set of team roles is not well balanced then the output will suffer badly, no matter how good the balance of functional roles. A team may be adequate in a functional sense, having the right expertise and experience, yet may not have the right balance of personalities to be productive. Belbin's (1981) research on management teams suggests that, to be productive, teams need a mix of personalities covering eight basic "team-roles," with the addition of a ninth ("specialist") role in technical situations (see also Stetter and Ullman, 1996). Using Belbin's (1981) terminology, together with descriptors to define them better in engineering design terms, these nine roles are:

- Company Worker   – practical organizer
- Chairman   – goal-setter and motivator
- Shaper   – dynamic pusher
- Plant   – creative problem-solver
- Resource Investigator   – information gatherer and negotiator
- Monitor-Evaluator   – option analyzer
- Team Worker   – perceptive listener
- Completer-Finisher   – conscientious perfectionist
- Specialist   – dedicated professional.

Belbin's set of team roles provides a simple way for a design manager to evaluate the strengths and weaknesses of a particular team, either by mental review or by the use of "Self-Perception Inventories" (Belbin, 1981) compiled by each person. The idea is not to search for nine individuals each filling one particular role, but to balance the team role distribution within the available team. The tendency to "hire in one's own image" needs to be resisted, as it can lead to a team strong in certain roles but weak in others. Individuals will often exhibit one dominant team role along with several secondary ones, and it is sometimes possible for the design manager to draw out secondary role traits to cover a weak area in the team, rather than bringing in an additional person. Positive contributions from team members are generally mixed with less useful ones, as indicated in Figure 3.1, and the team must be able to accept all the contributions, compensating for difficulties where necessary.

Ryssina and Koroleva (1984), in Russia, proposed almost identical team-roles, based on their study of team performance in engineering research institutes. They found that for teams involved in technological innovation the roles that were the key at any particular time depended on the phase of the project. The team-role balance that seems right for one phase of the design process may not be right for the next phase, and it is up to the design manager to monitor this and mold the team accordingly. For example, people full of conceptual ideas (Belbin's "Plant") are extremely useful at particular points in the design process, but for much of the design process the continual contribution of new ideas is a negative influence on project progress.

There are a number of other factors that have been found to be important with regard to design teams, and most of these are self-evident. There has to be cooperation, commitment, motivation, good morale, and effective communication, all of which are enhanced by the leadership of the design manager. The user may be involved as a member of the design team, sometimes to its benefit, but not necessarily so. Three of these additional factors deserve particular mention, these being the *negotiating ability* of the team, the *negotiating power* of the team, and the effectiveness of its *communication*. To be successful a design team needs to be good at negotiating, and it needs to negotiate from a position of power. In fact, design teams have more power than they often realize, because without their input almost any company would fail. The problem is that engineers are taught engineering, not politics, and they tend to be out-maneuvred when it comes to eloquent speeches. The more persuasively that design engineers are able to communicate their ideas and present their case, then the more control they will have over their design projects. Let us never forget the lessons to be learned from disasters such as the loss of the Space Shuttle Challenger. The inability of the design team to get their message across to a management concerned more with politics than detail design set the scene for a very public catastrophic engineering failure. If the design team had understood and learned how to use its latent power effectively, then Challenger would not have been launched. It is interesting to note that the world's great engineers, such as Eiffel and Brunel, were not only technically excellent, but were also persuasive, entertaining, and politically involved individuals.

TEAM ROLE	TYPICAL CHARACTERISTICS	POSITIVE CONTRIBUTIONS TO TEAM	NOTES AND OBSERVATIONS
COMPANY WORKER (Practical Organizer)	Conservative; dutiful; reliable; predictable	Organizes; turns ideas and plans into action through use of common sense and hard work	May tend to be inflexible and slow to respond to new possibilities
CHAIRMAN (Goal Setter and Motivator)	Mature; confident; trusting; controlled	Clarifies goals and priorities; encourages all to contribute; promotes decision-making	May take time to see potential of new concepts
SHAPER (Dynamic Pusher)	Highly strung; outgoing; dynamic	Challenges inertia; pressurizes; finds ways around obstacles	May tend to be impatient if progress is slow
PLANT (Creative Problem Solver)	Individualistic; imaginative; unorthodox	Creates original ideas; solves difficult problems	Unlikely to be good at communicating with others and managing a team
RESOURCE INVESTIGATOR (Information-Gatherer and Negotiator)	Extroverted; enthusiastic; curious; communicative	Explores new possibilities; develops contacts; negotiates	May tend to lose interest once initial enthusiasm has passed
MONITOR-EVALUATOR (Option Analyzer)	Sober; unemotional; intelligent; prudent; objective	Sees all options; analyzes; judges likely outcomes accurately; hard headed	Unlikely to motivate or inspire other team members
TEAM WORKER (Perceptive Listener)	Social; mild; accommodating; perceptive; sensitive	Listens; builds; averts friction; handles difficult people; promotes team spirit	May be indecisive at moments of crisis
COMPLETER-FINISHER (Conscientious Perfectionist)	Painstaking; orderly; conscientious; anxious	Searches out errors, omissions, and oversights; concentrates on and keeps others to schedules and targets	Inclined to worry about small things; may be reluctant to delegate
SPECIALIST (Dedicated Professional)	Professional; self-starting; dedicated	Provides specialized knowledge or technical skills	Contributes on only a narrow front

Figure 3.1. Summary of team roles (adapted from Belbin, 1981). *Courtesy of Dr R.M. Belbin*

Example: Gasifier Test Rig

To obtain some "team-role" data from the gasifier test-rig project, those participants contributing the most hours to the project effort were asked and encouraged to complete the "Self-Perception Inventory" as developed by Belbin (1981). Although the questionnaire was completed without adverse reaction by the contract staff, it was regarded with some suspicion by company staff, and the plans to gather such data for each phase of the project had to be abandoned. Only nine questionnaires were returned, of which only seven were complete! Despite the dubious response from the company staff, including a written commentary from one who felt that the questionnaire was biased in certain directions, the results were sufficient to indicate team-role differences between participants and the influence these had on the project:

- Contract staff had relatively even scores across all team roles, which indicated more of an ability to switch from role to role than to provide strength in one or two. The average score for all three of the contract staff showed most strength in the role of Company Worker and least in that of Monitor-Evaluator. The scores for two of these design engineers were virtually identical for six of the roles.
- Company staff scores showed more spread than those for contract staff, but the average scores for the group were uniform, as the highs and lows canceled out. The group appeared to be marginally stronger in the role of Plant over other roles, and slightly weaker in the roles of Company Worker and Completer-Finisher.
- The average scores for the seven Self-Perception Inventories varied very little from role to role, as the strengths shown by the scores of the contract staff tended to complement those shown by those of the company staff. This is somewhat academic, as two of the three contract staff were involved in the project only for short periods of time, but the project seemed to progress rapidly when these contract staff were present. This leads to speculation that they not only supported the team through functional roles, but also through an improvement in the overall balance of team roles.
- All three of the contract staff were professionally involved in design, yet their scores for the role of Plant (creative problem-solver) were lower than for most other roles. As the concept for the rig was considered satisfactory, this suggests that, for this project, the role of the creative problem-solver was less important than other roles.

3.4 Design Tools and Techniques

A large number of methods, tools, and techniques are now available for design, many of them on-line or computer-based. There are increasingly complicated codes, standards, manuals, and handbooks to assist the designer in specific areas. To help in understanding and applying all this in practice, there are numerous books and industry-based short courses. For example, the 2002 ASME Boiler and Pressure Vessel Code, published by the American Society of Mechanical Engineers, is comprised of 10 volumes and costs over US$8500 for the full set. It is available in printed form or on-line and the ASME can provide training courses, videos, books, and Code Committee assistance for users. For materials selection, software such as the Cambridge Engineering Selector (Cebon, 2003), with its standardized presentation of material properties, can help design teams to consider a far broader spectrum of materials than was practicable in the past. The 20-volume ASM Materials Handbook (ASM, 2003) is now available either in printed form or on-line for reference and provides detailed information on all aspects of materials use. It is not the intent to provide a list of tools and techniques here, but rather to encourage critical assessment of what is available and the judicious use of tools proven to be effective. Some useful suggestions for further reading are given in the Bibliography at the end of this book.

For drawings, calculations, modeling, simulation, and analysis the computer finally has become the indispensable design aid that it was supposed to be many years ago, but it is still true, as always, that it is a tool to be used and it is not a design engineer. Computers allow engineers more freedom, in that they can perform detailed analysis of more complex components. For example, finite element analysis (FEA) can be used quickly to predict the stresses, deflections, and natural vibration frequencies of components with complex geometries. Care must be exercised when using such methods, as computers are not intuitive devices. A novice FEA program user may not understand the limits of the software and the analysis can easily yield inaccurate results. In this case, the analysis should also be accompanied with a sanity check, such as the insight of a more experienced colleague, a simplified hand calculation, an experiment, or an independent check by a specialist FEA operator. On the other hand, it is easy to become constrained by "what can be done on the computer," and indeed to let the computer set criteria or churn out results that the designer-users cannot explain. For example, one of us was once assisting with the review of timing gear designs for two new truck engines, each engine being designed by a different team within the same company. There was a simple, but puzzling question: "Why do the timing gears for one engine have a helix angle of 15° whereas the others have a helix angle of 25.4043°?" The answer appeared to be that for one engine the helix angle of the gears had always been 15° and no-one had ever questioned it, whereas for the other engine the computer said it should be 25.4043° and no-one had ever questioned that either.

> ### Example: Formway Life Chair
>
> Computers were considered an essential tool for design of the Life chair at Formway: the Internet was used to gather design information, designers communicated with suppliers and external consultants using e-mail, and they worked collaboratively with designers at Knoll Inc. using file transfer protocol (FTP). Designers used this for design calculations, drawings, and project planning.
>
> The focus on environmentally sound principles in the design of the Life chair resulted in objectives such as minimizing the quantity of materials used to perform the same function, using nontoxic recyclable materials, and designing for maximum working life. Meeting these objectives required a detailed analysis of all the chair components. The team created 3D computer-aided design (CAD) models of all the chair components using SolidWorks®. The CAD models were converted into COSMOSWorks™ FEA models. The FEA models were then used to analyze the structural properties (predicting stresses and deflections) and also to model the flow of materials for the injection molding process. The design team did not include expert FEA personnel, so external consultants were commissioned to check the COSMOSWorks™ models. The results of strength and deflection predictions from external consultants were generally consistent with the analysis conducted by the Formway team. Following independent verification, physical models were produced using final materials and manufacturing processes.

Taguchi methods and the *quality function deployment (QFD) technique* (Clausing, 1994), developed in Japan to improve quality while reducing cost, can be helpful tools for the design engineer, especially for high-volume, mass-produced products such as motor vehicles. QFD provides a systematic means for identifying customer needs and translating them into a quantified design specification. For an introduction to this technique the reader is referred to the useful step-by-step summary with example presented in *The Mechanical Design Process* (Ullman, 1992). Genichi Taguchi's quality engineering system provides a systematic means for integrating quality engineering throughout the whole design process, rather than just being applied during particular phases. An example often used as an introduction to Taguchi methods (Ealey, 1988) concerns the closing of car doors. It was found that customers like the "feel" of a car door if it takes a force of about 3.6 kgf to close it. Japanese cars were found to close consistently at about this force, whereas American cars would either require a much higher force, or the required force would vary from car to car. The explanation for the difference is that the parts for

American car doors were considered acceptable if they fell anywhere within the specified tolerance limit, whereas the parts for Japanese car doors were made by aiming for minimum variability around *target specifications*. This meant that instead of the tolerance stack-up tending to result either in doors that leaked or doors that required a high closing force, the tolerance stack-up tended to result in doors that neither leaked nor required a high closing force. The customer perceives the more consistent product to be of higher quality, and the overall cost is lower as there is less need for expensive door-fit repairs on the assembly line. It is important to consider Taguchi methods in perspective with the overall design process, and to help with this the reader is referred to *Total Design: Integrated Methods for Successful Product Engineering* by Pugh (1990).

As an illustration of how the thinking about quality and the customer has changed since 1970, an engineer working for the Ford assembly plant in New Zealand at that time described how they used to receive all kinds of peculiar parts from the parent company in the UK, including leftover parts from earlier models. Shipments took so long to get to New Zealand by sea that it was assumed that the colonial Kiwis would find some way to build a car out of anything that arrived in the box, rather than return them to the UK and wait another 6 months for replacements. It was said that, on occasion, the last cars off the line would have two doors on one side and one on the other, with different door handles all round! No wonder that New Zealand customers would pay a premium to get an "English-assembled" car in those days, nor is it any wonder that almost every car on New Zealand's roads today is Japanese.

Communication is a critical factor in the engineering design process, and it is important to set up simple, reliable ways of exchanging information. Increasingly, the computer can be used effectively for this by means of e-mail, the Internet, and Web-based systems. Designers and managers need to communicate effectively with each other, the customer, and all those involved with the project. At the beginning, this means learning about the user needs, then negotiating and refining the design requirements to form a feasible set of project constraints and limits. The client must be kept informed of progress on a regular basis and should know the expected project outcomes. The iterative nature of the design process requires the continual revisiting of earlier work, based on the interpretation and processing of design notes, drawings, sketches, and calculations. It is important, therefore, that designers accurately record their ideas and analyses, whether working on their own or interacting with others. Communication channels must be kept open to allow questions and give early detection of problems with the design. In the project context, communication may be defined as *the controlled flow of design ideas and information*. The most effective means of communication varies from project to project. Figure 3.2 shows some common channels and communication interactions in a typical engineering project.

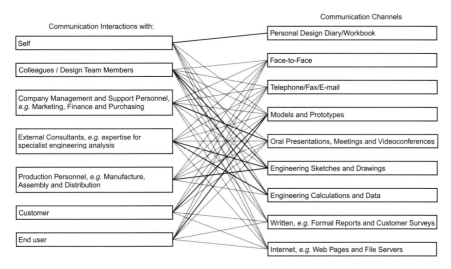

Figure 3.2. Some common communication channels and interactions in a typical engineering project

Engineering projects rarely end with the handing over of manufacturing documents. It is the designer's responsibility to retain complete and accurate records of all design information. This requires organizing the design information in clearly marked files. Projects often enter a monitoring phase once they enter the manufacture, assembly, testing, installation, and commissioning phases. A design on paper often accompanies a product through its service life and may be helpful in shortening the design process for future, similar products. In the unfortunate event of litigation involving a failure or dispute over product performance, it is essential that the design team is able to provide records of the agreed design requirement specification, the design calculations, and fulfillment of acceptance test requirements.

With regard to managing engineering projects, the checklists and work sheets provided at the end of each chapter in this book were designed specifically for use as a computer-based system, but this was impracticable with the available technology at the time of the first edition. Now that Web-based systems have become practicable, it has become possible for us to provide the whole set of checklists and work sheets on CD-ROM in a format that is suitable for the design manager to set up a simple Web-based design project review system to help in managing the engineering design process. The system is adaptable to different types of project and the sheets can be modified to suit the needs of a particular company. It may be used to provide a historical timeline record for reference on future projects, as well as to help the manager identify and address important issues on current projects.

Example: Gasifier Test Rig

The model of the engineering design process presented in *Engineering Design – A Systematic Approach* (Pahl and Beitz, 1984) may be taken as representing one of the more defined and detailed general procedures currently available to the design engineer and project manager. The use of these procedures during this project provided a structure for initially categorizing the field data describing all the design work done, and to allow a quantitative investigation of two particular aspects. These were the identification of phases and "steps" or activities in the engineering design process, and the use of design "methods and aids" or "design-related techniques." Only those project hours directly attributable to the engineering design process were counted for this part of the analysis. The input of management and others not involved in design work was excluded, leaving 27 participants with 2178 h (92%) of the total project effort.

The hours of each person were categorized firstly by phase of the design process and then by "step" within each phase, according to the Pahl and Beitz (1984) model. Much of the work effort could not be categorized in this way, and the following six additional activities were identified, not specific to particular phases (and, therefore, not "steps" in the Pahl and Beitz (1984) sense) but observed to occur in all phases.

General activities
- Planning work — personal day-by-day planning of activities.
- Reviewing/reporting — verbal or written project reports/reviews.
- Cost estimating — design costs, labor costs, hardware costs etc.
- Information retrieval — information processing of all kinds.
- Social contact — social interaction outside other categories.
- Helping others — informal help given on other projects.

The hours were also categorized by usage of design-related techniques as listed by Pahl and Beitz (1984). Again, much of the work effort did not fit any of these categories, and 13 additional techniques were identified. Those hours where no identified technique had been observed remained "Not classified." The additional techniques were as follows, grouped into three sets:

Working techniques
- Making lists — personal reminder lists.
- Cost estimating — all types of costing.
- Calculating — simple and complex calculations.
- Scheduling — use of bar charts, etc.
- Filing — making/using personal files of information.

Communicating techniques
- Questioning people — informal/formal, verbal/written.
- Presenting viewpoints — informal/formal, verbal/written.
- Negotiating agreements — informal/formal, verbal/written.
- Reviewing and reporting — informal/formal, verbal/written.

Motivating techniques
- Becoming involved — with the design, the person, or the situation.
- Injecting enthusiasm — conscious effort to raise level of enthusiasm.
- Adding humor — to break tension or bind group together, etc.
- Team building — conscious effort to optimize group effort.

Some of the quantitative findings may be summarized as follows:

- 47% of the engineering design effort could be categorized according to the Pahl and Beitz (1984) "steps" of the engineering design process.
- By adding six more "general activities" in each phase of the engineering design process, the remaining 53% of the engineering design effort could be accounted for.
- 22% of the observed engineering design effort could be categorized according to the "methods and aids" recommended by Pahl and Beitz (1984).
- Adding 13 more techniques for "working," "communicating," and "motivating" accounted for a further 74% of the total engineering design effort, leaving 4% unclassified.
- The activity that accounted for the most engineering design effort was found to be "reviewing and reporting," at 22% of the total.
- The design-related technique used most was "communicating by means of reviews and reports," observed as taking 15% of the total time.

3.5 Design Team Output

Most processes, even ones involving human activities such as the production process, may be analyzed in terms of measurable variables. A problem with the engineering design process is that so few of the many variables can be measured quantitatively, and in fact the only simple measure is work effort in hours. As the outputs from the engineering design process are less tangible than those, for example, from the production process or construction projects, percent completion is more difficult to estimate for design work, but it is still regarded as a useful measure of performance by management. The more the design process can be broken down into defined pieces of work, with some kind of tangible output at the end of each, the more realistically the time to completion can be estimated. Design reviews can be scheduled to assess the work progress in each phase (Alpert, 2003) and to plan ahead.

At the beginning of a design project it is common for only a few people to be involved on a full-time basis, and so it is easy to keep track of the work effort. However, it is a general characteristic of design projects that more and more people become involved as it progresses, and consequently it becomes more and more complicated both to keep track of hours spent and to predict what further hours are required. This is particularly the case during development of the concept and the detailing of every component after the conceptual design has been completed. The work tends to be lumped under the single title of *detail design*, and yet the number of hours spent on this is likely to be greater than all the rest of the project put together. For the purposes of *managing* engineering design, it is an advantage to break this down into *developing* the concept and *detailing* the concept, as there are tangible outputs from each and the very fact that the two things overlap with each other provides a means for assessing graphically how the project is doing and how the overall timescale could be shortened. For our management and analysis purposes, the design process has been broken into the four main phases of *Task Clarification, Conceptual Design, Embodiment Design*, and *Detail Design*, with an initial proposal or briefing phase termed *Project Proposal*.

If each person involved with a particular project simply records their hours worked, along with a two- or three-word description of the work done, then checks off which phase of the design process the hours best fit into, the design manager can build up a simple and valuable picture of design progress, both in terms of how much effort was required for each phase and percent completion of the overall design effort. As mentioned in Chapter 2, the unit of time should be in *hours*, not one-tenth of a hypothetical "average day" or "part of a week," or of any other variable unit set up by the company for other accounting purposes.

For example, consider the GTR project, which went through five phases of design as shown in Figure 3.3. It so happened that on this project the number of hours of work to complete each phase of the project was carefully recorded. A project proposal was prepared, submitted to management and approved for going ahead into preliminary design. The design specification was established and approved. Conceptual designs were generated and evaluated. The final concept was approved for development and experimental testing, and then this work was completed. Manufacturing processes and materials were considered and the detail design work carried out to finalize everything for manufacture.

If the information from Figure 3.3 is plotted on a phased timescale, then it can be mapped in the form of a *phase diagram* equivalent to that shown in Figure 3.4. In the first instance, let us assume that we know there will be a period of time between submission of the proposal and the start of the project and that the other project phases will overlap to a certain extent. It is also reasonable to assume that within each phase the design effort will build up to some kind of

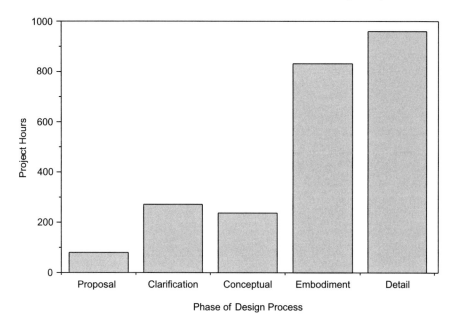

Figure 3.3. GTR project phases, showing effort per phase in hours

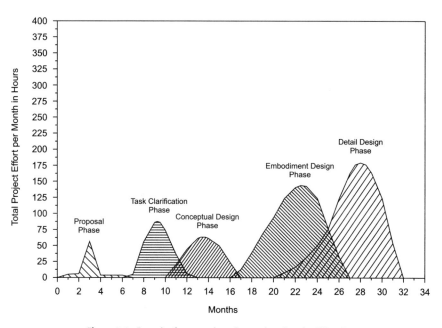

Figure 3.4. Example of a target phase diagram based on the GTR project

peak then die away again before the next phase starts. This particular diagram has a horizontal axis timescale in months and a vertical axis in hours of effort per month within each design-process phase. However, the axes can be varied to suit individual circumstances. For example, it may be more useful to plot percent of total project hours per month along the vertical axis. The phase diagram provides a means for visualizing approximately how much design effort will be needed in each phase of the design process and how much overlap can be achieved between phases. Based on this, a target graph of cumulative effort may be produced. Comparison of actual cumulative work effort against the ideal or target can then be used for monitoring and controlling of the design work based on achievable goals, and the design team will have a better chance of producing reliable estimates of "percent completion" and "cost-to-completion." This, in turn, gives the manager earlier warning of deviations and more time to take appropriate compensatory action.

To explore some of the features of a typical phase diagram, we will consider the actual phase diagram for the GTR project as shown in Figure 3.5. This shows the project effort in each phase by month and, therefore, indicates the measured overlap between phases. At a first glance the graph appears to have a lot of "noise," which would be increased if time was plotted in days or weeks and decreased if time was plotted in 2- or 6-month intervals. It became apparent,

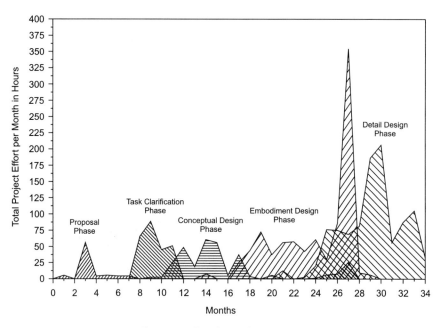

Figure 3.5. Phase diagram for GTR project

however, that the "spiky" nature of the graph plotted by month is significant for this particular project, as the major "peaks" and "dips" relate to specific events in the project history. Furthermore, it appeared that had such events not occurred, or had been foreseen and then compensated for, then the profile of phase-by-phase effort would have been more like that shown in Figure 3.4. In summary:

- If the project had gone exactly according to plan then the project phases would have been characterized by five bell-shaped curves, each overlapping others by a certain amount.
- In practice, the project did not go according to plan, and specific events caused specific "peaks" and "dips" in effort.
- Those dips caused by unplanned events reduced the proportion of work done within the envelope of the target phase diagram. For example in Figure 3.5, the major dip in effort for month 13 was caused by one team member's vacation, this at a time when the target phase diagram in Figure 3.4 would call for effort greater than that in month 12.
- As the design effort lost during the target time for a phase still had to be completed before the work of the next phase could proceed, for each dip within the envelope of a target curve there appeared a corresponding peak of effort to compensate, later in time and outside the target curve for that phase. For example, in Figure 3.5, it will be seen that to compensate for the dip in effort during month 13 a peak of additional effort occurred in month 17. This peak might have been expected in month 16, but the chance illness of a key team member delayed the work.
- Each such compensating peak delayed the finish time for that phase in the real case, diverting effort from the following phase and extending the overall project time.

Thus, Figure 3.5 is a useful summary of the overall project effort, and it can help to characterize the project. It shows that each project phase other than the proposal phase ended with a peak of effort apparently outside the target curve. Had the right things been done at the right time (*i.e.* effectively) and done in the best way when they were done (*i.e.* efficiently), then the work effort may well have been completed within the envelope of the ideal phase diagram and the project would have been completed sooner.

This suggests that the higher the peak-to-width ratio of each phase curve and the larger the overlap between phases then the more effective and efficient the project work effort would be; but this may not necessarily be so. For example, if embodiment design had overlapped with conceptual design then there would have been the risk that the "wrong" concept was being developed. On the other hand, once the layout of the simpler sub-systems had been agreed on through embodiment design there would have been an advantage in going straight on to detail design for those, which would have been indicated by greater overlap between the embodiment- and detail-design phase curves. Design activities also tend to benefit from an incubation time, particularly during the conceptual and embodiment phases, and a higher width-to-peak ratio may not allow sufficient time to develop ideas properly. It is not possible to generalize from such results, but it is certain that the flatter the curves, and the less the overlap, then the longer is the project time-span.

Figure 3.5 also shows another feature. It appears from the graph that the task clarification phase was completed in two distinct stages separated by a period of 18 months. In fact, what happened was that, owing to the effect of external influences, two changes were made in the design specification: the maximum operating pressure was increased from 102 bar (1500 psi) to 170 bar (2500 psi) and the maximum operating temperature increased from 1100°C to 1300°C. Unlike the late effort required to compensate for work not completed at the target point in time, this represents extra work outside the target envelope altogether. What the graphs cannot show is the extra work effort created in other phases by the change in specification during month 27; but, even ignoring this "knock-on" effect, it is clear that the additional task clarification effort added work hours and cost to the project. Design of the control system, which was completed almost as a "project-within-a-project" during embodiment design, also called for additional hours of task clarification.

At this point a number of questions might be asked, such as:

- How did project costs relate to project effort measured in hours?
- Did hourly charges reflect the relative "value" of project effort?
- What about wasted effort, mistakes, or mismatched expertise?
- What about people not always working to capacity?
- Were there "good" hours and "bad" hours in terms of outputs?

The only costs incurred during the project other than direct labour costs were incidentals such as traveling expenses, telephone charges, and postal charges. For the company staff these were included in the normal overhead added to the salary cost for in-house work, and for contract staff they were incorporated in the hourly charge rate used. This allowed the simplifying assumption to be made that project costs were proportional to project hourly charges. In addition, although there was a 3:1 ratio between the highest and lowest hourly charge rate, the recorded hours for the highest and lowest rates were so few by comparison with the total that they had little effect on the overall relationship between hours and cost. Thus, once the overall project cost had been calculated from the hours and cost-per-hour for each individual, a back-calculated average hourly charge rate gave a good overall approximation, and the project cost in pounds sterling could be considered directly proportional to project effort in hours. It also meant that although the "value" to the project of hours worked varied in a subjective sense, for the sake of quantitative argument, it could be reasonably assumed that all hours were of equal value. Simplification

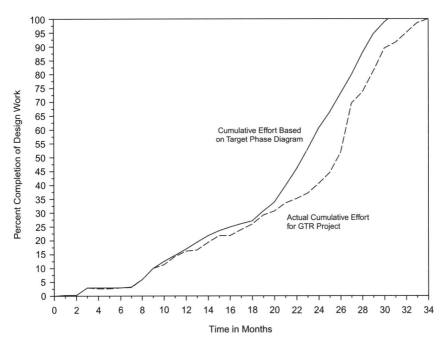

Figure 3.6. Assessment of percent completion for GTR project

is needed for analysis, and this can come through the use of the following two assumptions:

1. Project cost is directly proportional to project effort in hours.
2. All hours contribute equally to the project effort.

By plotting cumulative effort by time as shown in Figure 3.6, then by using the above assumptions it is possible to gain some idea of the *percent completion* at various points in the GTR project. The first 25% of project effort took 50% of the project time-span, and the first 50% of the project effort took 75% of the project time-span. Thus, 50% of the total project effort was completed in the final 25% of the project time-span. It is interesting to note that the 50% point in the project time-span was the point at which the conceptual design phase was ending and the embodiment design phase was beginning. This illustrates that, even for a project which did not have severe time constraints, most of the effort seemed to be put in at the end, and also that the use of company resources increased with time. The graph has the "S-curve" characteristics typical of graphs showing percent completion of construction and production projects, as described by Hajek (1977). A cumulative-cost graph would follow the same curve, closely matching the typical cumulative-cost S-curve suggested for use in engineering project cost control by Turner and Williams (1983) and

Darnell and Dale (1982). Comparison of the actual curve with the target curve provides a measure of where the project deviated from expectations, and by how much.

In order for the design manager to understand the full implications of the phase diagram for a particular project, the work progress must be assessed in context, and the way to do this is through regular design-review meetings (Alpert, 2003). What form these should take, who should attend them, and how often they should be held depends on the type of project and its features. The three main types of design project identified by Pahl and Beitz (1984) are:

- Original — starting from a blank sheet
- Adaptive — adaptation of a previous design for a new application
- Variant — variation to an existing design, such as a larger or smaller model.

And the features of projects within each type may be summarized in terms of the supplemented Rodwell (1971) scales described at the beginning of this chapter:

- Magnitude
- Complexity
- Novelty
- Production quantity
- Technical risk
- Urgency

It should be noted that the phase diagram for the gasifier test-rig project is particular to its project type and features, as would be any proposed target phase diagram. The target and actual phase diagrams for other project types and with different combinations of features may be quite different from those shown in Figures 3.4 and 3.5. Likewise, the type of production (job, batch, mass, or flow) will affect the design phases. The phase diagram technique is simply a useful way of presenting and analyzing the actual work effort in any design project against a proposed target, or against other projects for comparison purposes. Professor John Raine, and his colleagues at Canterbury University in New Zealand, have developed the technique further for different types of project. Their IMechE paper, "A study of design management in the telecommunications industry" (Whybrew *et al.*, 2002), shows the phase diagrams for the design of a heavy-duty "walkie-talkie"-type radio. This project, at Tait Electronics Limited, had a "design for manufacture" focus and CAD was used throughout, resulting in a phase diagram quite different than that for the gasifier test-rig project. The data-logging procedure and the application of phase diagrams provided a quantitative way of evaluating the complete design process, which in this case led to improvements with an estimated one-third reduction in time-to-market for later products.

Example: Product-integrity Board Meetings – Scott Fetzer

A large holding company with over 20 subsidiary manufacturers and an annual turnover in the billions of US dollars realized that they had problems with some product lines, and wanted to avoid future difficulties. A new approach to product design and quality improvement was instituted throughout the entire group of companies (Birmingham, 1991). For all new product lines, major design changes in existing products, or change in the use of products, *Product-integrity Board Meetings* are set up to review design quality, monitor design progress, and approve each phase of the design process. Each board is comprised of two corporate officers, management and design staff from the individual company, a senior engineer from another company, and an outside consultant. The board is responsible for ensuring that products are designed and can be manufactured to a quality, safety, and reliability standard that will meet or exceed customer expectations. At each meeting, the following aspects of the design project are reviewed and, depending on the phase of the design process, specific aspects are discussed in detail:

- Customer expectations
- Design specifications
- Test data
- User evaluation
- Risk analysis
- Invention analysis
- Process controls

The individual companies, or operating units, were generally wary of such procedures being introduced by the corporate headquarters, and there was reluctance to cooperate at first. However, the then corporate director in charge of quality improvement had degrees in industrial engineering, business, and law, along with an enthusiastic personality. His multidisciplinary background, coupled with persuasive skills and a friendly approach, overcame many initial problems, and one by one the operating units began realizing the benefits to be had from the Product-integrity Board Meetings.

PROJECT PROFILE CHECKLIST

LEVEL	INFLUENCES	CONTRIBUTING FACTORS	SOME QUESTIONS TO ASK: EFFECTS ON PROJECT?
PROJECT	DESIGN TASK	Magnitude Complexity Novelty Production quantity Technical risk Delivery time constraints	Is project too big/small? Is project too complex for time? Effect of novelty? Suitable for facilities? Within acceptable limits? Possible to complete on time?
	DESIGN TEAM	Expertise (competence) Experience Role balance Cooperation Commitment Motivation Morale Negotiating ability Negotiating power User involvement	Team competent to do work? Team experienced enough? Acceptable team balance? Are team members cooperative? Real commitment to project? Is team motivated? Effect of difficult times? Team able to meet needs? Team able to hold its own? Team/customer contact OK?
	DESIGN TOOLS	Systematic approach Formal design methods Intuitive design methods Communication Project control Computer design methods Computer aids Codes and standards	Effect of using systematic approach? Effect of using formal design methods? Effect of using intuitive design methods? Communication methods OK? Is there an effective system? Are these used to their capability? General facilities OK? Are relevant codes and standards incorporated?
	TEAM OUTPUT	Productivity Quality of work	Efficient use of time? Acceptable quality attained?

Figure 3.7. Project profile checklist

3.6 Project Profile Checklist and Work Sheet

To help the design manager quickly develop a realistic feel for the project being worked on, and to encourage the use of available tools and techniques, a *Project Profile Checklist* and *Project Profile Work Sheet* have been developed, as shown in Figures 3.7 and 3.8. A worked example based on the Life chair project is shown in Figure 3.9.

Contributing factors attributed to the *design-task* influence in the design of Formway's Life chair were as follows:

1. The *magnitude* of this project had a negative effect on the design team due to the level of detail at which designers were required to work. Designers were accustomed to managing the design of the complete product; however, this project required designers to work in groups on individual subassemblies and components. The magnitude was compensated for by

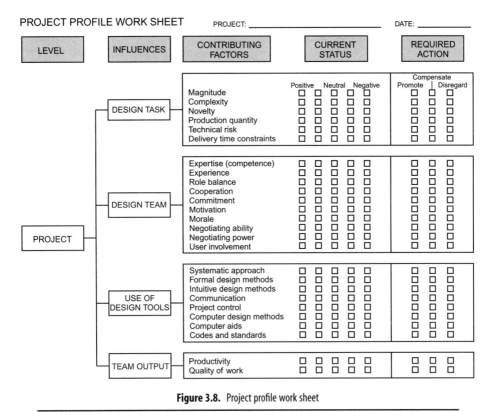

Figure 3.8. Project profile work sheet

detailed project planning and group coordination. For example, the current activities of each team member were listed on the project notice board so that team members could track the path of the project and perceive the "big picture."

2. *Complexity* and *technical risk* were considered to have negative effects on the design team. Although the team had previous experience in the design of office chairs, "Life" had a higher degree of functionality and was more technologically advanced than previous models. Setting delivery-time constraints that allowed the team sufficient time to deal with complexity issues and develop experience with new technology compensated for these factors.

3. The level of novelty for the chair project was high and had a negative influence on the project. This influence was offset by having a prototyping and model-making workshop attached to the design office, so that new technology could be tested and proven as the design progressed.

4. *Production quantity* had a negative influence because the Formway team were not experienced in, and did not have the facilities for, mass production. This was compensated for by drawing on the expertise and resources of their colleagues at Knoll Inc.

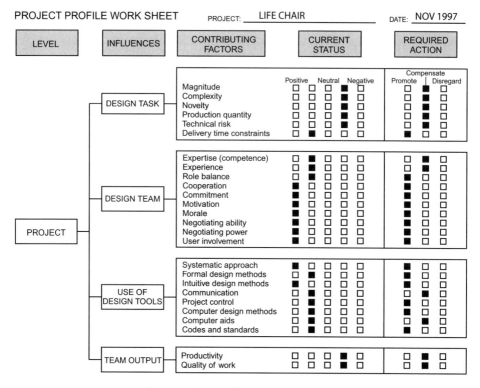

Figure 3.9. Example of project profile work sheet

The *design-team* influences (Figure 3.9) on the project were:

1. The design team had the necessary *expertise*, such as academic qualifications, and suitable work *experience* in the manufacture of office furniture, so these factors were considered positive contributions to the project. However, owing to some of the negative design-task influences for this particular project, the team realized that expertise and experience would need to be supplemented by employing external help.
2. The team was comprised of individuals with a range of suitable specialist skills, resulting in a good *functional role balance*. They had worked effectively on past projects and the composition of individual character traits (*team-role balance*) was considered likely to yield a productive group.
3. Team members were aware of their colleagues' strengths and weaknesses. The level of *cooperation* was high, and this contributed positively, allowing team members to work together through challenging tasks.
4. The transparency of company objectives, along with committed and enthusiastic management, resulted in a *committed* design team; this had a strong positive influence on the project.

5. The involvement of management who had confidence in the design team's ability, by giving praise for good work, resulted in a *motivated* design team.
6. The high level of complexity and novelty was recognized by management, and allowances were made in the project plan and in giving extra support, which held *morale* at a good level through difficult times.
7. The design team had the *negotiating ability* to acquire the project resources necessary to progress the project.
8. The *negotiating power* of the design team, within the company, was at a high level. This was because design was seen as critical to the company's success in the market.
9. *User involvement* contributed in a positive way, as customer trials created opportunities to establish the real needs of the customer. The team also became involved as users, by evaluating the prototypes and comparing these with competitor products.

The *design-tools* influences (Figure 3.9) on the project were:

1. Formway Design Studio adopted a *systematic approach*, with design activities and outputs closely resembling those shown in Figure 1.5. This approach had proven effective in the development other new and innovative office furniture products.
2. The team members were familiar with *formal design methods*, and these were considered to have a strong positive influence on the project.
3. *Intuitive design methods*, such as brainstorming, had a positive influence on the project. This was attributed to the experience of the design team and their high level of enthusiasm, which promoted the beneficial effect of such methods.
4. *Communication* was given a high priority and had a positive influence on the project. At Formway, the level of communication was high due to the close-knit design team and easy access to company management. At the detailed design and manufacturing phases, when a large US company was commissioned to manufacture the chair, there were perceived communication difficulties due to the geographically dispersed team. Communication was promoted by installing state-of-the-art information technology tools and by holding regular videoconferences at the detailed design stage. The US and New Zealand design teams found that their interpretation of the English language varied slightly due to cultural and social differences. Placing representatives from the US office in New Zealand and *vice versa* overcame interpretation difficulties during critical stages of the project.
5. *Project control* had a positive influence on the project. This was due to the effective systems in place to monitor the activities and progress of the small team.
6. *Computer design methods*, such as CAD and FEA, were used extensively throughout the project. Although these methods were essential tools and positively influenced the project, it was of recognized that a lack of resources

would need to be compensated for by employing the help of consultants with specialist skills.
7. *Computer-aids* positively influenced the project. The company was committed to maintaining a reliable computer network with up-to-date software tools.
8. Compliance with *codes and standards* was a goal that had to be achieved in the target market. *Codes* and *standards* contributed in a positive sense because these provided useful design information and set minimum levels of performance and safety.

In general, the *team-output* influences were negative because of the negative design-task factors of magnitude, complexity, and novelty, making it difficult to assess the project against team performance on previous projects. Negative *productivity* and *quality of work* factors were compensated for by allowing the team manager flexibility in matching the team size and level of expertise with the changing workload and technical demands. Negative *productivity* was compensated for by management allowing deadlines to be advanced when the original timescales were found to be unrealistic.

Influences at the design-task level were clearly not favorable for the chair project: resources were limited, and both the management and the design team were faced with an ambitious task (a higher level of magnitude, complexity, and novelty than ever before). Building a strong design team and providing adequate design tools and resources offset the negative effect of the task influence. The accuracy of the predicted team output was in doubt due to uncertainties associated with the type of task. This resulted in the project taking twice as long as originally planned. The project's success can be attributed, in part, to management's flexibility when team outputs did not meet original expectations.

3.7 Tips for Management

- Use a systematic design approach with phased activities.
- Develop a project profile to identify the staff, tools, and techniques needed.
- Negotiate wisely to ensure that there is sufficient money and time to do a quality job.
- Beware of unproven technology, increased complexity, and dangerously short timescales.
- Beware of large scale-up or scale-down factors from previous design work.
- Work out the ideal functional roles and team roles needed for each phase of the project.
- Match roles needed with the staff available for each phase of the project as closely as possible.
- Negotiate wisely to create the strongest design team possible.

- Use the most appropriate tools and techniques for the job in hand.
- Treat communication as an important management issue.
- Ensure that team members record their personal work on the design project.
- Ensure that appropriate communication channels are available.
- Establish a common platform where all parties can have their say.
- Introduce QFD and Taguchi methods where practicable.
- Use phase diagrams to help overlap phases and compress project timescales.
- Use phase diagrams as a means to assess percent completion.
- Develop review meetings or product-integrity boards for continuous design assessment.
- Use the checklist and work sheet to summarize status and establish a negotiating position.

Chapter 4
Managing the Design Team

4.1 Influences at the Personal Level
4.2 Knowledge, Skills, and Attitude
4.3 Motivation
4.4 Relationships
4.5 Personal Output
4.6 Personnel Profile Checklist and Work Sheet
4.7 Tips for Management

4.1 Influences at the Personal Level

Engineering design involves teamwork, and the better the team works the more likely it is that a high-quality design will emerge. Typically, design teams are multidisciplinary; and as the demands on the team change during the course of a project, so the composition of the team may need to be adjusted as the work progresses. Team members are now often "geographically dispersed," working far apart from each other, perhaps in different countries, cultures, and time zones. Communication has, therefore, become an even more critical issue. It is no longer simply a matter of making sure that information is sent and received, but ensuring that it has been interpreted and understood as originally intended. The use of computer systems for communication is no longer optional for commercial design projects – it has become essential for remaining competitive. The form of communication and the tools required may change depending on the design phase.

Many factors contribute to the success of a team, and in the literature it is suggested that an ideal engineering design team should be: competent; experienced; well balanced; cooperative; committed; and motivated! Some less well-documented factors identified are: morale level; negotiating ability; strength of power base within the company; end-user involvement; and the appropriate matching of design team composition to project requirements for each phase of the work. Although these factors may sound academic, in practice they are extremely important. We have assembled them into a checklist, with a corresponding work sheet, for evaluating the capability of a team working on a particular design project and the contribution of each team member individually. The following sections summarize the factors, grouped into areas of influence typically referred to in the literature.

> **Example: Formway Life Chair**
>
> In the early task clarification and conceptual design phases, the Formway team worked as a close-knit group within a small office. The team's workstations were arranged in groups corresponding to the chair sub-system designs (*e.g.* seating, support, and control), which promoted one-on-one discussion between team members. A team meeting/working area was created where the designers held daily project meetings and where team members could work in groups, trading ideas around a table or using a white board. The office was secure, so that project information could be posted on notice boards without worrying about confidentiality issues. During the later embodiment and detail design phases, the Formway team worked collaboratively with a similar team from Knoll. The resulting design team was then geographically dispersed and working in different time zones. This required new communication tools, and the team relied heavily on the use of daily videoconferences, e-mail, file-sharing software, and Web conferencing tools such as WebEx.

4.2 Knowledge, Skills, and Attitude

In common with many other intellectual activities, design work requires certain levels of *knowledge* and *skill* with an *attitude* of mind conducive to producing high-quality work in a team environment. It is sometimes suggested that creativity is the key to design; however, in practice, it is found that creativity is only part of the answer, and there are many important tasks that require systematic and detailed thought of a routine nature rather than design creativity. Does the person have a sound knowledge base for the work in hand? Is the person able to communicate ideas persuasively? Does the person work effectively and reliably to complete assigned tasks? Is the person's standard of work adequate for the particular project? Can the person prepare legible and neat handwritten notes? Can the person produce adequate sketches and drawings by hand? Does the person have a good attitude towards the management and the project? Does the person act as a good ambassador when meeting with customers or working with other companies?

It is important to assess such characteristics at the beginning of the project, as it may be very difficult to change the team once the project has started, and it is too risky to assume that there will be improvements in knowledge, skills, or attitude during the course of the project.

4.3 Motivation

It is the responsibility of the design manager to develop, encourage, and maintain a team that works well and produces high-quality results. This requires an abundance of *enthusiasm* and personal commitment on the part of the manager. Without it the project is likely to founder when the inevitable problems arise, such as shortage of time, money, or competent people. Enthusiasm motivates and helps to build up a reserve fund of goodwill for the future. Commitment, as used in the context of the design team, means full *involvement* with every aspect of the project and the *tenacity* or determination to see it through. It shows dedication to the project in hand, which is appreciated by higher management and puts the manager in a powerful negotiating position. It is very difficult to turn against someone who is obviously trying to do the best job possible for all those concerned, or to refuse reasonable requests with regard to resources.

For example, a manager may personally choose to work according to fixed hours of attendance, such as 8 a.m. to 5 p.m., or choose to work according to the needs of the project. The problem is that design projects do not necessarily progress well when confined to fixed hours in the day. If the manager's personal commitment to the project lies wholly within a rigid daily timetable, then why should it be different for anyone else in the team? It may be that there are some managers who are able to use every minute of every day so efficiently that they can rightly say that they are more productive than others who put extra time in; but this misses the point. The point is that if the manager is prepared to put in whatever it takes to make a success of the project, then this will be appreciated and will set the working tone for the team. A manager who works according to the project needs rather than by the time clock is also in a stronger position to deal with any personnel problems that should arise within the team.

Of course, there is another side to this question of commitment. Our experience has been that a dedicated design manager tends to become relied upon to do more and more, to the point of being totally overloaded, while others are not extended to their full capabilities. This state of affairs can result in a negative situation, which must somehow be controlled. A manager who "overdoes it" creates severe problems for everyone. It is necessary for the manager to develop an awareness of when to push the project forward with an additional burst of energy and when to let things progress at a more natural pace. Direct involvement with all aspects of the project is essential for building up credibility and respect within the team. It provides access to the inner workings of the team that helps the manager understand the technical and human issues well enough to deal with them competently at any particular level of detail. It also offers forewarning of any interpersonal problems that may cause severe setbacks. For example, the matter of some people working to the clock while others work to project needs is worth resolving immediately, or it may lead to unnecessary and unwanted friction between team members.

Frustration and *anxiety* are part and parcel of engineering design because of the risks taken and the uncertainty of the end result. For example, we have worked on many projects that started as high-priority jobs and which ended up being canceled for nontechnical reasons. This is frustrating for design-team members, but for the design manager it is far worse. What is the use of all the enthusiasm, involvement, and tenacity if the project is summarily canceled? The design team gets the impression that the manager is "crying wolf," and the people are not so easily motivated next time around. For the manager the project becomes an exercise in futility, and it is downright depressing. Often, few others care about the failure of the project, as a project, though the cost of any design work carried out may be quibbled over for a long time. Part of the reason for developing the design context checklist and work sheet in Chapter 2 was to provide the design manager with a means of assessing where a project stands in order to make a personal decision as to how to deal with the uncertainties identified.

Humor is a useful tool for a design manager, and may sometimes be used to advantage in defusing difficult situations or breaking a log-jam in project progress. However, it can easily backfire in serious situations, and it is best treated as a specific technique like any other, with times to be used and times not to be used. It is worth developing the ability to use humor effectively and in understanding differences in humor from one culture to another.

Example: Gasifier Test Rig

The month when almost twice the effort went into the project than in any other was when the contract controls engineer from Chicago temporarily joined the team. This engineer had not worked outside the USA before and was, therefore, operating in a foreign environment. However, he had both the expertise and experience needed for designing the control system, and the motivation and commitment to see this part of the project through. From the morning of Saturday 12 May, when he was met at Gatwick Airport by the contract design engineer, to the Saturday morning 2 weeks later when he flew back to Chicago, there was a marked change in the performance of the team. He was immediately accepted for the missing expertise and experience that he could provide, and for those 2 weeks he brought to the project a sense of purpose and urgency strong enough to ensure that the entire control system was designed within the 2 weeks. The process and instrumentation (P&I) diagram involving over

Continued

100 valves was completed; the seven control panels were detailed; sensor tables, valve operating sequences, and shutdown procedures were drawn up; a report was issued for use in the hazards analysis and in obtaining bids for construction; and a 2 h presentation meeting was held. Vacations were rearranged, a valve manufacturer offered enthusiastic help, management interest in the project was revived, and the project manager wrote to the contract controls engineer on 29 May: "The amount you accomplished in such a short time is beyond belief... it is very reassuring to have this essential part of it (the rig) defined with such skill and expertise."

4.4 Relationships

With so many team factors likely to affect a project, it appears that the design-team composition would be an important aspect, and there is evidence to support this. In Chapter 3, functional roles and team roles were discussed, and when it comes to the individuals themselves there has to be compatibility between team members, which will bind the team together rather than split it apart. As this is a dynamic situation, rather than a pre-existing condition, the design manager can do a lot towards encouraging and maintaining team-role compatibility. In fact, this may take a good proportion of the manager's time. As Sir John Egan once said, after becoming Chairman of Jaguar Cars Ltd: "It is amazing how little 10 000 people produce when they are all pulling in different directions, and it is even more amazing how much they produce when they are all pulling in the same direction."

Each person in the design team has relationships within the company and relationships outside the company. These relationships are important to the design manager, as they can greatly affect the productivity and work quality of the team. For example, a person who calls friends and family on the telephone for much of the day is unlikely to be adequately productive, yet maintaining happy family relations comes into the equation too. It is up to the design manager to be aware of what relationships exist and to deal effectively with those that show signs of being detrimental to the project.

4.5 Personal Output

People are individuals, with individual personalities and ways of working. From an engineering design point of view, important influences affecting the productivity, quality of work, power, and effectiveness of the design-team members may be summarized as follows:

- Enthusiasm
- Involvement
- Tenacity
- Compatibility

4.6 Personnel Profile Checklist and Work Sheet

To help the design manager in the difficult task of creating and maintaining an effective design team, a *Personnel Profile Checklist* and a *Personnel Profile Work Sheet* have been developed, as shown in Figures 4.1 and 4.2. For a small team of only three or four people, the questions in the checklist could be asked with reference to the team as a whole, and the work sheet used to record the assessment at the time. For larger teams, it is intended that the checklist and work sheet be used to assess individual contributions within the team. An example of a completed work sheet is shown in Figure 4.3, based on the team that designed the Life chair. As shown in Figure 4.3, factors attributed to the *knowledge* influence

PERSONNEL PROFILE CHECKLIST

LEVEL	INFLUENCES	CONTRIBUTING FACTORS	SOME QUESTIONS TO ASK: EFFECTS ON PROJECT?
PERSONNEL	KNOWLEDGE	Knowledge base Knowledge applicability	Are there knowledgeable specialists? Is knowledge matched to needs?
	SKILLS	Perception Use of knowledge Communication Creativity (imagination) Versatility Negotiation	Team decision-making OK? Are individuals knowledgeable? Team communications open? Sufficient? Too much? Adaptability in team? Team bargaining position?
	ATTITUDE	Work standards Self-discipline (habits) Integrity	Is the quality of work OK? Working habits OK? Are team members reliable?
	MOTIVATION	Enthusiasm Involvement Tenacity (determination) Frustration/anxiety Humor	Level of enthusiasm OK? Level of involvement OK? Are team members persistent? Effect of these on project? Do team members possess humor?
	RELATIONSHIPS	Team role compatibility Relationships within company Relationships outside company	Are team members compatible? Good relations between departments? Good relations with suppliers?
	OUTPUT	Productivity Quality of work	Is productivity of members good? Is work produced of high quality?

Fig. 4.1 Personnel profile checklist

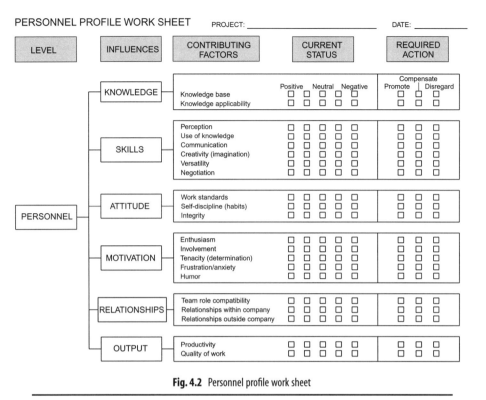

Fig. 4.2 Personnel profile work sheet

were considered positive. The team had designed three chairs prior to this project, and at the start of the project, their *knowledge base* was considered to be at a good level. At the later embodiment and detail design phases it was necessary to compensate for the *knowledge applicability* factor (due to the team's inexperience at designing for mass production) by seeking the help of external experts. By the end of the project, the Formway team had learnt the skills required for developing mass-produced office products, and hence they were in a better position to succeed at this type of work in the future.

Factors affecting the *skill* influence (Figure 4.3) at a personnel level were:

1. *Perception* had a positive effect at a personal level: individual team members were experienced and perceptive; however, team decision-making suffered initially due to the dominance of some individuals within the team. This matter was resolved with the help of external team management consultants, who introduced team rules for meetings, such as no criticism of ideas when generating concepts, allowing equal participation, and by ensuring all team members have the same level of input during project meetings. The result was a team with effective decision-making capability.

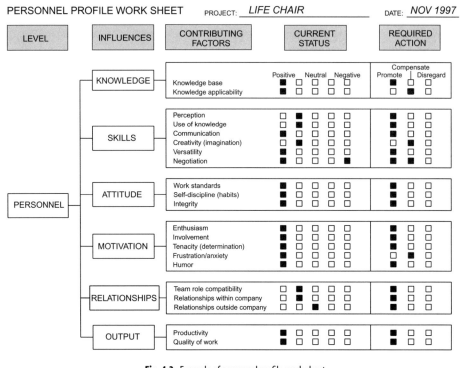

Fig. 4.3 Example of personnel profile work sheet

2. The *use of knowledge* factor was positive; this was attributed to a good blend of individual skills, such as practical model-making, specialist ergonomics knowledge, and professional engineering expertise.
3. The team had excellent *communication* skills and this was promoted, positively influencing the project.
4. *Creativity* was considered neutral overall, even though the project team was highly creative. They were experienced at creativity techniques, such as brainstorming, and they were proficient at building concept models and mock-ups. This resulted in 12 patents and an innovative final solution for the chair. However, the positive effect of this highly creative capacity was offset because the team found it difficult to stop the creative process when the time came to concentrate on detail design.
5. The Formway team was *versatile* owing to multi-skilled individual team members and a flexible management team. These factors had a strong positive influence.
6. The Formway team had excellent *negotiating skills*, and this, coupled with management perceiving the project to be of high importance, enabled the team to get necessary resources within the company. Externally, the Formway team's negotiating ability with management at Knoll was poor. This was due to the fact that the Life chair was just one product in a huge

range. Maintaining a firm position with open communication channels compensated for weaknesses in external negotiating skills.

The *attitude* influences (Figure 4.3) on the project were:

1. Consistently high *work standards* within the Formway team. This was attributed to a high level of ownership and a quality of the work environment.
2. *Self-discipline*, which was considered to have a strongly positive influence. Close communication within the design team meant that self-discipline issues were identified early and systems were in place so that these could be resolved quickly.
3. *Integrity*, which was strongly positive due to clear role descriptions and personal targets with a responsive management team.

Factors contributing to the *motivation* influence (Figure 4.3) were:

1. There was a good level of *enthusiasm* towards the project, and this was attributed to the team members sharing a sense of worth and value within the company. Their contribution was valued and this promoted.
2. The whole team was *involved* in project activities such as meetings and workshops. The team also had representatives at top-level management meetings. This had a positive effect on the project.
3. A good level of *tenacity* existed within the team. This was attributed to individuals having a strong sense of ownership of the project as a whole and having the freedom to pursue their own ideas.
4. Team members had adequate training and tools available, which reduced the likelihood of *frustration*. When major problems did occur, these were identified quickly and turned into a positive. This minimized the effect of *anxiety*. In the event of a major failure, the team would put the event in a positive light, often simply laughing about it and then getting on with fixing the problem.
5. *Humor* was considered an essential ingredient in the design process. Included in the project mission statement were the words "make it fun."

Factors contributing to the *relationships* influence (Figure 4.3) were:

1. *Team-role compatibility* was considered critical to the success of this project. The current responsibilities of each team member were posted on a team notice board, ensuring clarity of roles. A high emphasis was placed on quick resolution of conflict. In the event that conflict or team roles could not be resolved, then team members would be removed and assigned to another project.
2. *Relationships within the company* were promoted by ensuring open communication channels. This had a positive effect because the team members were able to acquire necessary resources and support for their project working.
3. *Relationships outside the company* were promoted by ensuring that external personnel were informed of project developments, by maintaining personal-level relationships, and by holding regular project reviews.

Factors contributing to the *output* influence (Figure 4.3) were:

1. *Productivity* was considered a function of the team output rather than individuals. It was important for productivity to be at a good level, and this was promoted by providing the team with a supportive environment and access to necessary resources.
2. *Quality of the work* was of a high standard, and this was credited to matching skills with tasks.

From the completed personnel work sheet for the Life chair, Figure 4.3, it is evident that the Formway design team's experience could be applied to develop a new seating concept. This experience included relevant work skills and strengths such as communication, versatility, and negotiation, which would allow them to embrace new technologies and the tools needed to move into unfamiliar areas. The strengths of this team were undoubtedly their professional attitude and motivation. This resulted in highly productive work output of excellent quality.

4.7 Tips for Management

- People are individuals, with individual personalities and ways of working.
- Each person comes with certain knowledge, certain skills, and a certain attitude.
- The design manager is in a position to motivate each person, and to encourage:
    Enthusiasm for the project;
    Involvement with the project;
    Tenacity or persistence in getting the job done;
    Compatibility and cooperation with others on the project.

PART 3
The Project

5   Project Proposal: Getting the Job
6   Design Specification: Clarification of the Task
7   Feasible Concept: Conceptual Design
8   Developed Concept: Embodiment Design
9   Final Design: Detail Design for Manufacture
10 Users and Customers: Design Feedback
11 Standards and Codes
12 Engineering Design Process: Review and Analysis

Chapter 5
Project Proposal: Getting the Job

5.1 Proposals and Briefs
5.2 Preparing a Proposal
5.3 Negotiations
5.4 Debriefing
5.5 Project Proposal Checklist and Work Sheet
5.6 Tips for Management

5.1 Proposals and Briefs

At the start of most new projects enthusiasm runs high, the coffers are assumed full, and hard engineering is a long way down the road. Projects start in many different ways, depending on circumstances. Often a design project stems from an idea or an identified need in a market, and competitive bids are solicited to help develop the idea or solve the problem. Such a call for bids may vary from a simple verbal request up to a complicated written *request for proposal* (RFP) in several volumes, depending on the organization and the scope of the work involved. The response to this is a *project proposal* or *bid*. As the primary aim of a project proposal is to secure some form of contract, the proposal is generally a written document formally stating what the bidder understands the project to involve and the approach proposed to satisfy the request. In the first instance it is used for bid selection and evaluation, then later for negotiating terms of contract with the successful contender.

A project may also start in response to a request for specific design work, presented in the form of a *project brief*. There is no set format for project briefs; they range from simple verbal requests to the presentation of detailed documents defining the work to be carried out. Whereas a project proposal is written with the aim of securing a contract, the main aim of a project brief is to define the work to be carried out. This is more the equivalent of the *work statement* section of a project proposal than the overall document. Within a project brief, or instead of it, there could be a more narrowly defined *design brief*. Briefs may also be used as working documents during the course of a project. For instance, a design brief might evolve into a design specification. There is often confusion surrounding the use of the term "brief." This arises because different versions of the same document may be used, for example, as the "request for proposal," the "proposal" itself, and the "design specification" during a project. Misunderstandings are likely to arise unless it is made absolutely clear which version of the brief refers to which aspect of the negotiations.

5.2 Preparing a Proposal

The first step in the business of securing a contract by means of a proposal is to obtain the RFP, or the bidding documents, as early as possible. This takes time, effort, and planning, just like any phase of the design process. There is also some strategy involved, as indicated in Figure 5.1. Just looking for advertisements, such as in the *Commerce Business Daily* (see Bibliography), is a weak starting point. The idea is to reduce one's "handicap" progressively by making contacts, learning about what is going on in the field, checking on what the competition is up to, and developing a solid reputation in the area for high-quality design and high-quality products. This may go so far as having input into the RFP in the first place.

Then comes the critical decision on whether to bid or not to bid for a particular contract. It costs money to bid, and the chances of winning may be remote. Well-prepared input from the design manager is essential at this point. A wide range of factors must be considered, such as the type of project, scope of project, availability of resources, and the predicted return on investment. The

BIDDER STATUS	WAYS OF QUALIFYING
SOLE-SOURCE	The only supplier, or by far the best
INVITED BID	Past performance and reputation
BIDDER'S LIST	Proven capability with recognized expertise
RFP DISTRIBUTION	Credibility and enthusiasm in the market area
ADVERTISEMENT	Accepted capability in the market area
UNSOLICITED PROPOSAL	Submission of project suggestion or funding request

Figure 5.1. Hierarchy in bidding process

technical and financial risks involved must be evaluated, and even the time and expense of preparing the proposal itself may be a significant factor. One of us once led a proposal team bidding on the design of diagnostic probes for a fusion power reactor. Although the chances of winning were remote and the cost was high, it was decided to submit a proposal on the basis that the design team would gain knowledge, information, and experience in a completely new area. This could then be used to submit further proposals with greater chances of winning a contract as we became progressively more in tune with the customer's needs. All those bidding were invited to visit the fusion reactor facility and to attend debriefing sessions after the contract had been awarded. In fact, our three-volume proposal (technical, management, and pricing) was ranked high in the final assessment order, which moved us up to "invited bid" status for the next round of proposals.

During planning of the proposal effort it is important to consider a series of issues based on the three mentioned in the Introduction: *design activities*, *design outputs* and *influences*. Who will be in the design team? How will the team have to change throughout the course of the project? Is the team capable of producing a finished design of high enough quality? What are the other influences likely to affect the project? Will their effect on the project be positive or negative? Which are fixed and which are variable? Which ones can the design team control?

Many useful guidelines for the preparation and presentation of project proposals have been published. For example, Hajek (1977) covers those for large one-off projects in considerable depth, and Warby (1984) deals with smaller-scale projects. The most important guideline for proposal preparation is to *be responsive*. It is no use preparing a bid that does not meet the customer's requirements. The project will go to someone more attuned to that customer's needs.

Proposals are usually assessed according to criteria drawn from the RFP document, and irrelevant material gains no credit. RFPs usually contain specific instructions as to the content and presentation, but, if not, the following is a useful guide as to structure and content:

- Summary of complete proposal.
- Background and qualifications of organization ("boilerplate").
- Statement of the problem as understood.
- Technical discussion to show understanding of the problem and how solutions will be found.
- Statement of work to be carried out.
- Management of project, including organization, resources, key personnel, and team output.
- Project plan to show phases, timescales, design review meetings, and decision points.
- Estimated costs, suitably itemized by time, phase, or deliverable.
- Concluding statement to sum up the high points of the proposal.
- Supporting appendices, only if necessary and only if allowed.

Once the design problem has been appropriately formulated it is likely that the vague initial ideas on how the project should be structured, scheduled, and managed can be set down more formally. This is a necessary part of preparing a proposal, and the earlier it is done the better, even if there is not enough information to do it accurately. It is only then that the real constraints of time, human resources and financial resources become apparent. Reporting procedures can be set, preliminary tasks allocated and approximate time schedules estimated, together with design-team requirements. Often, all that is necessary is a simple Gantt chart or bar chart such as the one developed for the gasifier test-rig project, as shown in Figure 5.2. For large projects, planning techniques such as PERT are recommended to establish initial priorities, but it must be realized that updating such planning charts can itself become a costly and time-consuming exercise unless kept very simple. Low-cost computer packages have now simplified the business of formalizing project planning charts and are well worth using, even on small projects. There are a number of texts useful in planning projects listed in the References and the Bibliography; in particular, see Hajek (1977), Leech and Turner (1985) and Turner and Williams (1983).

Example: Gasifier Test Rig

On 18 May a project proposal was submitted to the company outlining the design approach, together with a cost estimate and a project plan. This proposal was accepted on 2 August and the design effort started on 1 October. The task was to design a high-pressure, high-temperature materials test rig. Although the main needs for the rig had been identified, it was seen as having several likely uses, and the requirements were thus "ill-defined." No design specification (requirements list) existed. A series of rigs had been constructed and operated by the same project team, so that this project was seen as another in a progressing sequence; but, as this rig would involve the difficult problem of handling flowing coal at high temperature and pressure, the design task was considered to be both novel and complex.

The project team initially included two managers, one section leader, two research scientists, one design engineer, and the researcher (Hales), as a "participant observer." As was normal practice on such projects, no person was assigned to it full time, so everyone had other responsibilities. It was agreed at the beginning that the project team should be flexible and that specialist help would be called on as required. During initial meetings the project team stressed that they had no structured approach to the design of special-purpose equipment and were keen to develop one. The interest was in an integrated procedure rather than in merely the application of certain techniques, so the decision was taken to structure the project according to the systematic approach then being developed at Cambridge University.

Project Proposal: Getting the Job 99

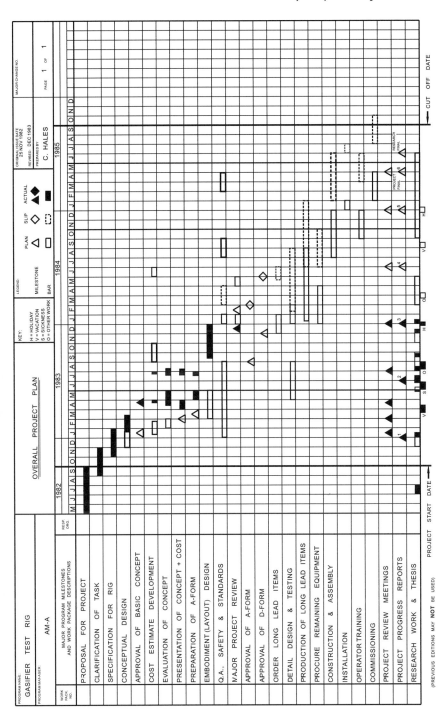

Figure 5.2. Overall project plan for GTR project

5.3 Negotiations

Producing a proposal always seems to require burning the midnight oil, with many aspects to bring together, complicated formats to comply with and other work coming to a halt as the deadline approaches. There is a sigh of relief as the Federal Express van speeds away and work gets back to normal. However, unfortunately there is still more to do after delivery of the proposal, and very often it is left unfinished once previous priorities resume.

First of all, the files of papers, correspondence, drawings and other materials used for preparing the proposal need to be reviewed and reorganized, ready for answering questions that might be asked during evaluation of the proposal. The questions are likely to be specific and probing, possibly catching the design manager unawares if poorly prepared. If the files are well organized and any further thinking has been recorded then it is much easier to develop responsive answers in a timely manner.

Secondly, the project team needs to prepare for oral presentations and negotiations in the event that the proposal reaches the final selection stage. Weak points in the proposal must be addressed, information regarding competitor bids should be analyzed, and a strategy developed for how best to present the features of the proposal in a convincing and persuasive way. There may be negotiations on price, performance, scope of work and many contractual details. The design manager needs to be clear as to the negotiation boundaries set by upper management, and to make sure that those involved with the negotiations fully understand the technical risks and uncertainties inherent in the proposed project. Here again, if the project has been planned based on use of a systematic design approach, then it will be possible to define phases and the likely work effort required within each phase far more accurately than if the whole design process is treated as a single entity.

There is, of course, a sense of elation if yours is the winning proposal and you are awarded a contract. This is a time when all of a sudden upper management seems to be much in evidence, darting around in excitement. The design manager may get forgotten in the rush for the drinks, or alternatively may suddenly become the focus of attention. Either way, experience would indicate that the wisest thing to do under the circumstances is to "remain cautiously optimistic" and make sure that the bill for the dinner is not charged to the project! One of us was once enjoying an unforgettable meal with upper management in a downtown Chicago restaurant on the day of a contract award, when even the effects of the wine could not blur the image of the first 2000 dollars of the project money disappearing without a trace. A year later the same upper management was demanding to know why the project was going to overrun. From a design point of view the dinner is best left until "it works."

5.4 Debriefing

There is no guarantee that a contract will be awarded, even for an excellent proposal that seems to meet all the customer's requirements. It can be disheartening to learn that, despite all your best efforts, somebody else got the job and you are $10 000 out of pocket. However, every proposal is a learning experience and the thing to do is gain the most from it. Attend the debriefing session if it is offered and ask for one if it is not. Find out what features the winning proposal had over yours and why yours was considered inferior. It may be possible to get a copy of the winning proposal to use as a model for future reference, and it may be possible to work with the company or organization that was awarded the contract. Sometimes a company is unable to meet the commitments made in its proposal and the sponsor or customer has to start negotiating again. Submitting a revised proposal might be in order in such a case. Nothing is cast in stone, and it pays to follow up quietly but actively to see what transpires and what possibilities may arise in the future.

5.5 Project Proposal Checklist and Work Sheet

The *Project Proposal Checklist* and *Project Proposal Work Sheet* shown in Figures 5.3 and 5.4 have been developed to help the design manager cover important issues with regard to proposals. By reviewing the checklist questions and then using the work sheet, the design manager will automatically check on the proposal process at the same time as compiling a current status report on the proposal preparation. It enables the setting of priorities and assignment of tasks to be carried out in an organized and rapid way. An example of a completed work sheet, based on the Life chair project reconstruction, is shown in Figure 5.5. The Formway team researched the needs of potential end users thoroughly in order to formulate a comprehensive brief for the Life chair project. From this brief, and their past experience in furniture design and manufacture, they were confident they had the skills and resources to complete the project. The design team prepared a proposal, which was presented to management for approval to start the project. This proposal included a clear problem statement, and identified key personnel and resources available both within the company and from external sources. A comprehensive project plan and indicative costing were included in the proposal; however, the work sheet shows that confidence was marginal, and this was due to their unfamiliarity with projects on a mass-production scale. The team realized that the project plan and costing would need continual revision until they had a better understanding of the magnitude of the task.

The design team was confident in negotiations with management because management had shown confidence in their ability in the past. Representatives from management were involved in the preparation of the proposal, which helped the team to match the proposal with the management vision for the company.

5.6 Tips for Management

- Aim to improve bidding status progressively.
- Get in early and get to know the customer's needs.
- For a winning proposal, be responsive.
- Prepare for questions on weak areas.
- Prepare in advance for oral presentations.
- Clarify negotiation boundaries before meetings.
- Request and attend debriefing sessions.
- Hold the dinner at the end of the project.

PROJECT PROPOSAL CHECKLIST

PHASE	ISSUES TO CONSIDER
PROPOSALS AND BRIEFS	Can you do the job? Are you qualified to do the job? Have you the resources to do the job? Are you registered on the bidder's list? Have you received all the RFP documentation? Have you enough information to prepare a proposal? What influences are likely to affect the project at different levels of resolution? Are you confident you can handle the influences? Have you planned how to go about doing the project? Do you want the project?
PREPARATION OF PROPOSAL	Executive summary included? Background and qualifications of organization included? Statement of problem included? Technical discussion clear, concise and accurate? Statement of work included? Organization described? Resources compatible with project? Experience and qualifications of key personnel included? Project plan included? Estimate of cost included? Conclusion adequately supported? Appendices necessary and sufficient? Proposal responsive to request for proposal? Proposal submitted on time?
NEGOTIATIONS	Prepared to respond to weaknesses in proposal? Prepared for oral presentation? Project plan included? Update accounting system? In a position to start work if contract is awarded? Prepared for final negotiations?
DEBRIEFING	Who won the contract and why? What was better about the winning proposal? What were the weaknesses in your proposal? What other factors influenced the decision?

Figure 5.3. Project proposal checklist

PROJECT PROPOSAL WORK SHEET

PROJECT: _____ DATE: _____

PHASE	CONTRIBUTING FACTORS	CURRENT STATUS (Good / Marginal / Poor)	REQUIRED ACTION (Proceed / Revise / N/A)
PROPOSALS AND BRIEFS	Competent team Qualified team Adequate resources Registered on bidder's list Received RFP Sufficient information Likely influences Confidence level Planning OK Project wanted	☐ ☐ ☐ ☐ ☐	☐ ☐ ☐
PREPARATION OF PROPOSAL	Executive summary Boilerplate Statement of problem Technical discussion Statement of work Organization structure Resources description Key personnel Project plan Cost estimate Conclusions Appendices Responsive to RFP Proposal on schedule	☐ ☐ ☐ ☐ ☐	☐ ☐ ☐
NEGOTIATIONS	Weaknesses addressed Oral presentation Project planning Accounting system Team organization Proposal analysis	☐ ☐ ☐ ☐ ☐	☐ ☐ ☐
DEBRIEFING	Overall proposal assessment Possibility for future bids Likelihood of winning in future	☐ ☐ ☐ ☐ ☐	☐ ☐ ☐

Figure 5.4. Project proposal work sheet

Project Proposal: Getting the Job   105

PROJECT PROPOSAL WORK SHEET   PROJECT: _LIFE CHAIR_   DATE: _NOV 1997_

PHASE	CONTRIBUTING FACTORS	CURRENT STATUS					REQUIRED ACTION		
		Good		Marginal		Poor	Proceed	Revise	N/A
PROPOSALS AND BRIEFS	Competent team	■	□	□	□	□	■	□	□
	Qualified team	■	□	□	□	□	■	□	□
	Adequate resources	□	■	□	□	□	■	□	□
	Registered on bidder's list	□	□	□	□	□	□	□	■
	Received RFP	■	□	□	□	□	■	□	□
	Sufficient information	□	■	□	□	□	■	□	□
	Likely influences	□	■	□	□	□	■	□	□
	Confidence level	□	■	□	□	□	■	□	□
	Planning OK	■	□	□	□	□	□	■	□
	Project wanted	■	□	□	□	□	■	□	□
PREPARATION OF PROPOSAL	Executive summary	■	□	□	□	□	■	□	□
	Boilerplate	□	□	□	□	□	□	□	■
	Statement of problem	■	□	□	□	□	■	□	□
	Technical discussion	■	□	□	□	□	■	□	□
	Statement of work	■	□	□	□	□	■	□	□
	Organization structure	■	□	□	□	□	■	□	□
	Resources description	■	□	□	□	□	■	□	□
	Key personnel	□	■	□	□	□	■	□	□
	Project plan	□	□	■	□	□	□	■	□
	Cost estimate	□	□	■	□	□	□	■	□
	Conclusions	□	■	□	□	□	■	□	□
	Appendices	■	□	□	□	□	■	□	□
	Responsive to RFP	■	□	□	□	□	■	□	□
	Proposal on schedule	□	■	□	□	□	■	□	□
NEGOTIATIONS	Weaknesses addressed	■	□	□	□	□	■	□	□
	Oral presentation	■	□	□	□	□	■	□	□
	Project planning	□	■	□	□	□	■	□	□
	Accounting system	□	■	□	□	□	■	□	□
	Team organization	■	□	□	□	□	■	□	□
	Proposal analysis	■	□	□	□	□	■	□	□
DEBRIEFING	Overall proposal assessment	■	□	□	□	□	■	□	□
	Possibility for future bids	□	□	□	□	□	□	□	■
	Likelihood of winning in future	□	□	□	□	□	□	□	■

Figure 5.5. Example of project proposal work sheet

Chapter 6
Design Specification: Clarification of the Task

6.1 Problem Statement and Design Specification
6.2 Defining the Problem
6.3 Project Planning
6.4 Demands and Wishes
6.5 Design Specification
6.6 Design Specification Checklist and Work Sheet
6.7 Tips for Management

6.1 Problem Statement and Design Specification

In order to carry out a design project successfully, two things need to be established as early as possible:

- A clear statement of the problem to be solved, for which solutions will be sought.
- A set of requirements and constraints against which to evaluate the proposed solutions.

The first is termed a "definition of the problem" or *problem statement*, and the second is termed a "specification," a "target specification," or perhaps more accurately a *design specification*. Both of these are essential if a solution to the problem is to be found that satisfies all parties. Considerable effort, and possibly some preliminary design work, may be needed to help establish what the real problem is, but it must be done. Finding a solution to the wrong problem is unacceptable design practice. Similarly, if the design specification inadequately defines the requirements and constraints, or contains ambiguities, then inevitably there will be clarifications required later at additional cost. It may not be possible to finalize every detail of the design specification at this stage of the project, but the process of preparing and obtaining approval for the design specification will itself help to identify specific items left unresolved. Allowance can be made for coming back to them later.

The QFD technique mentioned in Chapter 3 is one tool available specifically for helping to identify the real needs of customers and users, establishing what problem is to be solved, and developing a quantitative design specification as a basis for the design work that follows (Clausing, 1994). Other approaches, such as that of Cagan and Vogel (2002) or Ulrich and Eppinger (2004), may be more appropriate, depending on the project.

6.2 Defining the Problem

The natural tendency to accept a problem as given and to begin thinking of solutions must be resisted or the wrong problem may be addressed. Sufficient time must be spent in clarifying the task. It is good practice to define the problem in writing as a first step. This is not as easy as it sounds, for design problems are rarely what they appear to be at first sight and they may change with time. Design problems are not the same as analytical problems, although analytical approaches are often used during the course of design work. For example, some characteristic differences between design problems and analytical problems are shown in Figure 6.1.

	PROBLEM TYPE	
	ENGINEERING ANALYSIS	ENGINEERING DESIGN
PROBLEM AREA	Clearly defined	Poorly defined
STATEMENT OF PROBLEM	Precise	Vague
INFORMATION AVAILABLE	Sufficient	Insufficient
FINAL SOLUTION	Single solution	Compromise of many solutions

Figure 6.1. Engineering analysis compared with engineering design

In general, a design problem summarizes what is undesirable in a particular situation, and the problem is considered solved when an improvement in that situation is achieved that is acceptable to all parties. This will be a compromise solution as distinct from a "correct" solution. It is often helpful to try and formulate the problem at a higher level of abstraction (more generally) than first stated, and Pahl and Beitz (1984) offer the following questions to be asked in doing this:

- What is the task really about?
- What implicit wishes and expectations are involved?
- Do the suggested constraints actually exist?
- What paths are open for development?
- What properties must the solution have?
- What properties must the solution not have?

> **Example: Texas A&M University**
>
> It is an unforgettable experience to attend a game at the Texas A&M University football stadium, not only because of the unusual traditions involved, but also because of some interesting design features of the stadium itself. The stadium is built to cater for some 70 000 enthusiastic supporters. Consider the design problem: *How to provide drinks for 70 000 people during a football game?* One can imagine obvious solutions, such as to bring in cans of soft drink and sell them from stalls and perhaps the problem would be solved. However, a little more thought reveals that any such solution immediately creates a huge secondary problem, *i.e.* cleaning up after the game. This is not mentioned in the above problem statement. However, the most surprising thing about a football game at Texas A&M is the absence of litter afterwards. The 70 000 people leave and the stadium is clean! The secondary problem has somehow been addressed.
>
> Now consider the combined design problem: *How to provide drinks for 70 000 people during a football game without resulting in a litter problem afterwards?* This is a different problem and it will result in different solutions. In the case of Texas A&M University the solution was to serve drinks in simple, but well-designed, plastic cups inscribed with the "Texas Aggies" logo of the football team. Every piece of potential litter was converted into a prized souvenir, immediately collected up by youngsters eager to do a little business on the side.

6.3 Project Planning

Concurrently with the formulation of the problem statement, and usually as part of the design specification, the timescale and engineering effort required for the design must be estimated. Considering the wealth of existing design experience, it is surprising how frequently the design time for a project is still underestimated, often by a factor of two or three times. A likely reason for this is that so few historical project records are kept which break down the design process into clearly defined phases and itemize the real cost in terms of work hours. The tendency is just to lump the whole design effort together and guess a total cost. But this will inevitably lead to large errors. A better way is to break the design part of the project down into phases appropriate to the specific project, then itemize design tasks in as much detail as possible within each phase. As discussed in Chapter 3, if phase diagrams are available from previous projects then this provides an excellent starting point for planning the next project in a realistic and quantitative fashion.

> ### Example: Formway Life Chair
>
> A Gantt chart was prepared at the start of the Life chair project. This included detailed predictions of the times relating to each of the project activities within each design phase. Despite this, the project overran on time by a factor of two. This was attributed to the fact that the design team and its management support were working at a production level that they were unfamiliar with. They underestimated the level of detailing required to produce a product for mass production as opposed to batch production. They had no historical data that could be used to help in predicting the times for each activity.

There are many good texts on project planning, some of which are listed in the Bibliography and on the CD-ROM at the end of the book. They provide a spectrum of techniques appropriate to different types of design project, including computer-based approaches. All that needs to be done here is to emphasize the importance of project planning and to recommend, as a first step, the compiling of a Gantt chart similar to the one already shown in Figure 5.2.

6.4 Demands and Wishes

When the design problem has been appropriately formulated, it is helpful to compile a list of *demands* and *wishes*, or essential and preferred requirements regarding potential solutions to the problem, as described by Pahl and Beitz (1984, 1996). This is a simple way of compiling the information necessary to develop a well-structured design specification.

- *Demands* (D) must be met at all times, or the proposed solution is not acceptable.
- *Wishes* (W) should be taken into account, but only within acceptable costs.

Demands and wishes should be quantified whenever possible, and it is sometimes helpful to rank their importance. As an example of the difference between demands and wishes, consider the Texas A&M problem again. There may have been a *Demand* that all litter associated with the distribution of drinks be eliminated, with a *Wish* that all litter of any description be eliminated. It is then possible to break the *Wish* down further by, for example, ranking plastic litter as a higher priority for elimination than paper litter, and so on.

6.5 Design Specification

A design specification is a formal working document compiled principally from the list of demands and wishes or more sophisticated techniques such as QFD. It is an indispensable lifeline for the design engineer, and it should be treated in a serious fashion as follows:

- Insist that sufficient time is allotted to produce a comprehensive design specification.
- Divide the document into sections, such as shown in the Design Specification Checklist at the end of this chapter.
- Itemize each contribution and identify it as a *Demand* or *Wish* (or equivalent).
- Circulate to *all* involved so that everyone is formally consulted and can contribute.
- Identify the source of each requirement by logging the contributor's name against each item.
- Date any changes and identify the contributor requesting the change.
- Update the document during the course of the project to provide an accurate working record.

There are various types of specification, such as task specification, product design specification and material specification, as discussed in Chapter 31 of *Design Management: A Handbook of Issues and Methods* (Oakley, 1990). A task specification lays down what is required from the design team for a particular project (*i.e.* activity), whereas a product design specification (Pugh, 1990) lays down the requirements for the product to be designed (*i.e.* object). Here, we are particularly concerned with the product design specification, or design specification for short. Once the design problem has been defined and the requirements have been listed in the form of a design specification, a firm base has been established for the project to proceed through the conceptual, embodiment and detail design phases. Solutions to the defined problem may be sought, and the resulting concepts may be evaluated against the design specification. It is important that the design specification does not specify predetermined solutions to the problem or include "fictitious constraints", which might limit the design concepts to known, but non-optimal solutions, as shown by the vertical lift unit example in Hales and Pattin (2002).

The systematic preparation of a design specification, recording the source of every requirement, is highly recommended as it helps to avoid many potential later disagreements. However, it is not always appropriate or necessary to go about compiling a design specification in such a formal way. When design engineers are very close to the requirements of a particular type of product, a written specification may not exist and the team will work according to more or less an *implied* design specification. In this case, because of the designers' long-term involvement with previous similar products, they will know what the product's functional requirements are, how it will be manufactured, and how it will meet the needs of their customers. The risk here is that, immediately the design team deviates from known types of product, the implied specification will be inadequate. This was made painfully clear in the international arbitration concerning the design of the two huge custom-built tri-axis transfer presses mentioned as an example in Chapter 3. The company manufacturing the presses in one country was requested by the customer to have the design of the machines carried out by a specific company in another country so as to incorporate "new technology" through a licensing agreement. This resulted in a high-risk situation where the design of the machines was almost totally separated from manufacture, and for machines that were larger and more complex than either company had produced before. Both companies were experienced in press design and in press manufacture, and they had worked together before. This was reflected in the design specification for the two new machines, which was patched together from previous contract specifications and was full of implied requirements. It appeared that such requirements were normally assumed in the conventional press business. However, this particular design called for combining conventional press lines with an automatic tri-axis transfer feed system that raised the complexity level of the overall machine. Although all parties agreed to the design specification, it did not define the machine performance and controls adequately. This created major misunderstandings and ambiguities, which resulted in a massive lawsuit.

It is important that the design specification is circulated to all parties involved in the marketing, design, manufacture and distribution of the product, as well as to the customer (if appropriate). Once all parties have approved the final design specification, any changes requested should be treated formally and not simply agreed to as a matter of course. Any change that is not the correction of an error is likely to have cost repercussions, and often a change in one place affects the design somewhere else, not necessarily in an obvious way. The person requesting the change should justify the change, in writing, and the change should be incorporated into an updated version of the design specification together with the signature of the requester and the date of the change. If this procedure is followed then only those changes felt to be of real importance will be made, and the line of responsibility for the effects of such changes will be clear.

> ### Example: Gasifier Test Rig
>
> The task, as defined by the company and formally set down in a project proposal, was to design a materials test rig to operate under particular high-pressure, high-temperature conditions. Although the main needs for the rig had been identified, it was seen as having several likely uses and the requirements were thus "ill-defined." No design specification existed. A series of rigs had been constructed and operated by the same project team, so that this project was seen as another in a progressing sequence; but, as this rig would involve the difficult problem of handling flowing coal at high temperature and pressure, the design task was considered to be both novel and complex.
>
> A problem statement was prepared, and by questioning all project participants according to the checklist shown in Pahl and Beitz (1984) a comprehensive list of demands and wishes was generated. The detailed design specification compiled from this was a 20-page document listing 308 requirements and constraints, of which 217 were demands and 91 were wishes. Thirteen people contributed directly, and 34 requirements came from the 400 ideas generated by a brainstorming session involving 15 people. The requirements fell into four categories: function, production, operation and general information. Each requirement was labeled with the name of the contributor, and the document was circulated to all the project participants for review and modification by a set date. A total of 92 corrections, clarifications, and additions were made, involving 72 demands and 20 wishes. Once all parties had agreed on the specification, only two items were changed during the rest of the project, and these were caused by specific external influences. Details are given in Hales (1987).

The procedure suggested by Pahl and Beitz (1984) was used for the gasifier test-rig project and was regarded as so effective by the project team that it was adopted for use on other projects. Previously, researchers needing a test rig would sketch out the requirements in the form of a concept, and submit this either to the senior design engineer in the company or to an outside supplier. Design work would begin, and there would often be misunderstandings and problems, leading to disagreements and wasted effort. A major reason for this was the lack of involvement of groups such as safety specialists at the task clarification stage. Important requirements would be omitted from the initial list, and continual changes would be made during the rest of the project. Figure 6.2 shows that, for the gasifier test rig, over 40% of the design requirements came from sources other than research staff and, in particular, 19% came from

SOURCE	FUNCTION OF RIG	PRODUCTION OF RIG	OPERATION OF RIG	INFORMATION FOR DESIGN	TOTALS BY SOURCE
RESEARCH STAFF	77	48	36	21	182
MANAGEMENT STAFF	7	4	7	7	25
SERVICES STAFF	4	18	9	27	58
OTHER SOURCES	18	10	3	12	43
TOTALS	106	80	55	67	308

Figure 6.2. Breakdown of GTR design specification by source and type of requirement

the services staff responsible for manufacture. The procedure used for this particular design specification almost doubled the list of requirements that might have been expected had normal company practice prevailed, and it ensured that a comprehensive set of criteria was prepared for the selection and evaluation of conceptual solutions to the design problem. It also forestalled a number of later difficulties in the project. Of the 533 work hours spent on task clarification and conceptual design, preparation of the specification took 170 h, or 32%.

6.6 Design Specification Checklist and Work Sheet

By combining the list of demands and wishes (or equivalent) with the results of project planning, a comprehensive design specification can be compiled to provide the maximum design freedom within the given constraints. This is best structured in tabular form according to a checklist. Starting with the one offered by Pahl and Beitz (1984), for example, a simple but comprehensive *Design Specification Checklist* has been developed as shown in Figure 6.3. It is strongly recommended that design specifications should be compiled on a set of standardized sheets by use of a computer. As a starting point, the *Design Specification Work Sheet* shown in Figure 6.4 has been compiled, and a blank work sheet is also included in PDF format on the CD-ROM inside the back cover of this book. The way this is used is to fill out a set of work sheets for each category of item checked off in the boxes at the top of the page. For each category shown there is a set of questions in the checklist, and the idea is for this checklist to prompt sufficient questions that a comprehensive design specification can be compiled for review and comment extremely rapidly. A comprehensive design requirement specification was formulated for the Formway chair project. Figure 6.5 shows a sample page, which includes general "ecological" requirements for materials selection. Eco-design principles were developed by Formway in collaboration with RMIT University and were considered to be key to the success of this product. The design specification list included over 50 ecological requirements.

Other techniques for preparing a design specification, using similar checklists and resulting in similar formats, will be found in the material listed in the Bibliography.

6.7 Tips for Management

- Define the problem in words or symbols.
- Use a questioning checklist or QFD to itemize demands and wishes (or equivalent).
- Ensure that a realistic project plan is prepared, acceptable to all parties.
- Compile a detailed design specification with input from all people involved.
- Label all design specification items with the name of the contributor.
- Circulate the design specification to all those involved, for comment and approval.
- List all changes and additions with the date and the name of contributor.
- Obtain formal approval for the document in writing.

DESIGN SPECIFICATION CHECKLIST

REQUIREMENTS	CONTRIBUTING FACTORS	POINTS TO CONSIDER
FUNCTIONAL	Overall geometry Motion of parts Forces involved Energy needed Materials to be used Control system Information flow	Size, height, width, length, diameter, space, number, arrangement Type, direction of motion, velocities, acceleration, kinematics Load direction, magnitude, weight, load, impact, stiffness, inertia Heating, cooling, conversion, efficiency, pressure, temperature, storage Flow, transport, properties, implications, regulation, life-cycle Electrical, electronic, hydraulic, pneumatic, mechanical Inputs, outputs, form, display, computer
SAFETY	Operational Human Environmental	Direct, indirect, hazard elimination, safeguarding Warnings, training, instruction, personal protection Land, sea, air, noise, light, radiation, reaction, transport, emergencies
QUALITY	Quality assurance Quality control Reliability	Regulations, standards, codes, accreditation Inspection, testing, measuring tolerances, labeling Design life, failures, statistics
MANUFACTURING	Production of components Purchase of components Assembly Transport	Factory limitations, maximum dimensions, means of production, wastage Supplier quality and reliability, inspection Special regulations, installation, siting, foundations, bolting, welding Material handling, clearance, packaging
TIMING	Design schedule Development schedule Production schedule Delivery schedule	Project planning, project control Design detailing, in-house tests, compliance tests Manufacture, assembly, quality assurance, packing, transport Delivery date, distribution network
ECONOMIC	Marketing analysis Design costs Development costs Manufacturing costs Distribution costs	Size of market, strength of market, distribution, servicing Design team, computing, information retrieval, reproduction Design detailing, supplier costs, testing costs Tooling, labor, overhead, assembly, inspection, cost to customer Packing, transport, service centers, spare parts, warranty
ERGONOMIC	User needs Ergonomic design Cybernetic design	Type of operation, instructions, warnings Human interface relationships, operation, height, layout, comfort, lighting Controls, layout, clarity, interactions
ECOLOGICAL	Material selection Working fluid selection	Solid, liquid, gas, stability, protection, toxicity, safety Liquid, gas, flammability, toxicity
AESTHETIC	Customer appeal Fashion Future expectations	Shape, color, texture, form, feel, smell Culture, history, trends Rate of change, trends
LIFE-CYCLE	Distribution Operation Maintenance Disposal	Means of transport, nature and conditions of dispatch, rules, regulations Quietness, wear, special uses, working environments, foreseeable misuse Servicing intervals, inspection, exchange and repair, painting, cleaning Recycle, scrap

Figure 6.3. Design specification checklist

DESIGN SPECIFICATION WORK SHEET

Name of Project:	Issue Date:	Page: ___ of ___

Requirements:
- ☐ Functional
- ☐ Safety
- ☐ Quality
- ☐ Manufacturing
- ☐ Timing
- ☐ Economic
- ☐ Ergonomic
- ☐ Ecological
- ☐ Aesthetic
- ☐ Life-cycle
- ☐ Other

Demand/Wish	Itemized List:	Name of Contributor:	Date of Change:

Figure 6.4. Design specification work sheet

DESIGN SPECIFICATION WORK SHEET

Name of Project: *FORMWAY OFFICE CHAIR "LIFE"*	Issue Date: *February 1998*	Page: 21 of 30

Requirements:
- ☐ Functional
- ☐ Safety
- ☐ Quality
- ☐ Manufacturing
- ☐ Timing
- ☐ Economic
- ☐ Ergonomic
- ■ Ecological
- ☐ Aesthetic
- ☐ Life-cycle
- ☐ Other

Demand/Wish	Itemized List:	Name of Contributor:	Date of Change:
	Materials selection		
D	Material quantities minimized without compromising function, quality, aesthetics, or applicable standards	J.G., K.V. & M.P	
W	Materials with recycled content (post-consumer)	J.G., K.V. & M.P	
W	Materials void of toxic/hazardous substances	J.G., K.V. & M.P	
W	Materials derived from renewable sources	J.G., K.V. & M.P	
D	Materials commonly recycled and supported by collection systems	J.G., K.V. & M.P	
W	Materials produced using low-energy methods	J.G., K.V. & M.P	
D	Materials must not contribute to Sick Building Syndrome or other indoor air quality problems	J.G., K.V. & M.P	
W	Materials that are non ozone depleting	J.G., K.V. & M.P	
D	Minimize diversity of material types used	J.G., K.V. & M.P	
D	Wood-based materials and natural fibers from sustainable agricultural operations sustainably managed plantations and certified accordingly	J.G., K.V. & M.P	
W	Avoid wood-based materials containing toxic or hazardous substances (attn to urea formaldehyde)	J.G., K.V. & M.P	
D	Use textiles that are woven or dyed through cleaner production methods	J.G., K.V. & M.P	
D	Allow for refurbishment and recycling when specifying textures and designing fastening methods	J.G., K.V. & M.P	
D	Minimize off-cuts & by products & other materials wastage	J.G., K.V. & M.P	
W	Eliminate use of solvent-based adhesives and finishes	J.G., K.V. & M.P	
W	Eliminate use of finishes that contain heavy metals	J.G., K.V. & M.P	
W	Use materials with sensorial properties that positively contribute to healthy workspace (color, texture, surface design)	J.G., K.V. & M.P	
W	Specify durable materials avoiding colors that will date	J.G., K.V. & M.P	
D	Include relevant symbols for recycling	J.G., K.V. & M.P	
W	Parts should break down to discrete material types to reduce material contamination when recycling	J.G., K.V. & M.P	

Figure 6.5. Example of a design specification work sheet. Courtesy of Formway Design

Chapter 7
Feasible Concept: Conceptual Design

7.1 Divergent and Convergent Thinking
7.2 Generating Ideas
7.3 Selecting and Evaluating Concepts
7.4 Estimating Costs
7.5 Presenting the Final Concept
7.6 Conceptual Design Checklist and Work Sheet
7.7 Tips for Management

7.1 Divergent and Convergent Thinking

Once a design problem has been defined and a formal design specification developed there is sufficient information for a concentrated effort on conceptual design. Perhaps ideas have already been collected from earlier thinking or "inventions" have begun to surface. This can all be put to good use in generating the most ideas possible in the time available, then selecting and evaluating them to determine the most promising candidates for development. A large number of techniques are available for generating and handling ideas, starting with those described decades ago by Jones (1970), and it is up to the design manager to orchestrate their use in the most effective way.

Often it is possible to consider the overall device or system in a holistic manner, and sometimes the most suitable solution will emerge from this, particularly if the required functions are simple. For more complex systems, however, a systematic approach, such as that offered by Pahl and Beitz (1984, 1996), can improve the overall yield of ideas. This involves the breaking down of the overall problem into sub-problems, finding solutions to each sub-problem, and then combining them to form overall solutions, as was done with the Life chair (Figure 7.1). An advantage of this more systematic approach is that it involves recording the process and the likelihood of good solutions remaining undiscovered is diminished.

During the previous phase of clarifying the task and developing the design specification, the aim was to take a wide range of information and condense it down into essential and desirable features that any design solution should have. This *convergent* thinking is essential for preparing a concisely defined problem with a comprehensive set of requirements. When it comes to the conceptual design phase, however, the aim is different, and a different mode of

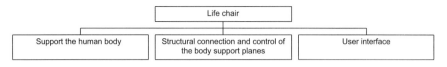

Figure 7.1. Breaking down the overall problem

thinking is required. Starting with the defined design problem or *problem statement*, the aim is first to generate as many ways as possible of solving the problem, then to select the most promising ideas that meet the *design specification*, then to evaluate them and determine the most appropriate solution to the problem, termed the *concept*. There are specific changes in thinking from *convergent* during task clarification to *divergent* while generating conceptual solutions and then back to *convergent* again during the formal selection and evaluation of the most promising conceptual designs. This is shown diagrammatically in Figure 7.2.

Conceptual design involves a combination of both types of thinking and, although they might be applied in a haphazard way in practice, it is important for the design manager to understand clearly which is being used when. It is then possible to steer the design team towards a definite end point in a short space of time while providing the most conducive environment for fostering creative solutions. Being innovative for the sake of being innovative is risky and unnecessary, especially when there is a good existing solution to a problem. On the other hand, an innovative solution can often lead to a winning design.

Divergent thinking means broadening out to collect as many ideas as possible, using the many different established techniques such as brainstorming, checklists, and "morphology" charts. Convergent thinking means weeding out the weaker ideas and homing in on those with the most promise, again using the many different established techniques such as selection charts and decision matrices. It is easy to get into a muddle, mixing divergent and convergent think-

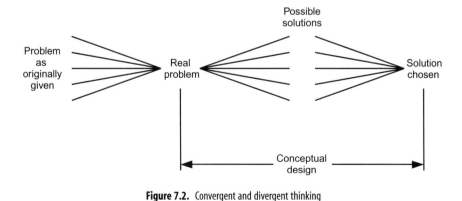

Figure 7.2. Convergent and divergent thinking

ing at the same time, or for one person to be in a diverging mode while another is in a converging mode. For example, a cutting remark from one team member about an idea being put forward by another will often kill the idea before it has had a chance to germinate. It is up to the design manager to recognize this common difficulty and to protect fragile new ideas from early obliteration. Another difficulty is in knowing when to call a halt to diverging thinking and to organize the less-exciting business of extracting the best ideas in a somewhat rational manner. "Ideas" people, or those who consider themselves "creative," will tend to keep producing more solutions or pressing hard for the adoption of a favorite. If all else fails it might be necessary to divert their attention on to other work in order not to end up with an argument instead of the most viable and practical solution.

> ### Example: Corkscrew and Winebox
>
> The first patent for a corkscrew was taken out by a Reverend Samuel Henshall in 1795, and during the 19th century there were several hundred more patents and registered designs filed. Today, the corkscrew is available in a huge variety of configurations, the most sophisticated of which incorporate lever arms to eliminate holding a resistant bottle between your feet, and protective bushings to prevent damage to the neck of the bottle.
>
> Despite all these developments, however, the concept remains unchanged. The corkscrew is a means for getting a cork out of a bottle, and the more the merrier. Why do we want to get corks out of wine bottles? So as to pour the wine out into glasses for drinking. How many glasses of wine do you get from a bottle? How many corks do you have to pull for a party? What a slow procedure, unless it is part of a fine dining experience. It is said that it was an Australian who made the mental leap, perhaps during a party: "Geez another wine mate, I'm dry as old bones." Talk about goatskins of water in the desert? What about wine flasks? What about wine casks? Hey, what about wine in a plastic bag with a tap on it, shoved in a box for rigidity? The winebox was born!

Although it may not be proven scientifically that humans have different thinking abilities in the left-hand and right-hand sides of the brain, from an engineering design standpoint the model is helpful. Let us accept for a moment that right-hand brain strengths are in creative, divergent thinking and that left-hand brain strengths are in logical, convergent thinking, and that sketching, describing and discussing, for example, are external means to assist in communication between the two halves of the brain. In terms of this model, the orchestration of conceptual design can be seen very simply. For generating ideas, encourage right-side thinking and team up with those who seem to have a right-side brain bias. Encourage *externalization* by all different means to maximize communication, both within individuals and amongst team members. For selection and evaluation of concepts, make a conscious switch to encourage left-side thinking and team up with those who seem more adept at this.

The aim is to muster all available right-side thinking into the most concentrated effort possible, then do the same with left-side thinking. If only one person is involved in this then the likelihood is that the right-side thinking, the left-side thinking, or both will be at a mediocre level and the final concept will reflect this. Engineering education currently tends to concentrate on the development of left-side thinking almost to the exclusion of the other, and this is a major problem when it comes to design. The design manager must be keenly aware of the creative limitations of the design team and ensure that all sorts of other people, at many levels, are drawn in to augment the right-side thinking, at least for a short period of time (*e.g.* in a carefully organized brainstorming session). It is worth thinking about the difference between a *designer* and an *inventor* at this point. An inventor comes up with ideas that may or may not be worth pursuing, and every now and then the chances are that a viable idea will surface. In some cases it becomes a winner. A design engineer defines a technical problem based on a set of requirements and sets off to find the most appropriate solution to the problem within defined constraints of time, money, and other resources. In no way can the design manager rely on the fact that a good idea might show up at the right time. It is absolutely critical that a sound concept is developed within the time allotted. The most reliable way of ensuring this is to generate as many good ideas as possible, then beat them around to determine which one stands up the best. Fixing on one concept without properly considering alternatives is often attractive and seems expedient at the time, but it can set the course for disaster later. This was part of the problem with the Space Shuttle Challenger.

7.2 Generating Ideas

From the foregoing it might appear that the ideal would be for the whole team to engage in a single bout of furious idea generation, followed by a sober round of decision making, but this is highly unlikely in practice. It is much more likely that there will be several iterations in the process, and there may be a mixture of complete solutions and partial solutions to evaluate. Some effort is required by the design manager to bring all the alternatives to about the same level of development and detail before any serious evaluation takes place. One structured procedure for generating ideas and producing alternative arrangements is that suggested by Pahl and Beitz (1984), summarized as the following set of guidelines:

- Abstract the problem: broaden it out and make it more general to understand the real issues.

- Formulate the overall function: describe what the thing is supposed to do.
- Break down into sub-functions: describe the different functions necessary for the thing to work.
- Draw up a system flow diagram (function structure): diagrammatically summarize the above.
- Generate ideas and concepts using selected creative methods: refer to available techniques.
- Determine different solution principles for each sub-function: all the different ways it can be done.
- Combine solution principles to generate complete solutions: overall concepts.

Example: New Concepts to Support the Human Body – Life Chair

An extensive ergonomic requirements specification list was developed, similar in form to the ecological requirements list example shown in Figure 6.5. Abstracting the essential requirements from this list, and considering the "support the human body" sub-problem of the three described earlier in this chapter, led the team to the crux of the problem, which was identified as: *support the human body without constraining it*. The design team realized that the human body was not designed to sit still; it was designed to move. Their goal, therefore, should be to develop a support system that promoted continuous natural movement. For design purposes the human body support function was divided into four sub-functions: seating; lumbar support; upper back support; and arm rests.

The Formway design team worked together utilizing the team's wide range of experience to brainstorm ideas for each sub-system, then smaller groups were formed to capitalize on the strengths of individuals to develop these ideas further in specialist areas. For example, a small group of designers worked on the seat sub-system. This involved developing concepts for the seat pan and it's support structure. Initial ideas were developed using hand sketches and simple prototype mock-ups. Prototypes were then built to prove novel ideas. "Support without constraint" required improvements in both comfort and circulation. The team worked on reducing the likelihood of pressure points behind the knee, a common problem associated with existing office chair designs. Their idea was to create a "waterfall" effect by incorporating a very flexible leading edge on the seat pan. Figure 7.3 shows an early prototype, where a flat plastic seat pan was cut along the posterior–anterior direction to allow greater flexibility.

Concepts were developed to accommodate the user's ischial bones. Considerations here were to reduce pressure points and to stop the user

Continued

from sliding forward in their seat. Figure 7.4 shows an early prototype developed for this purpose.

The concepts shown in Figures 7.3 and 7.4 were combined into working-solution variants, and full working prototypes were built and evaluated.

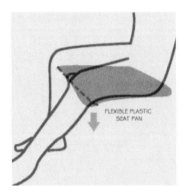 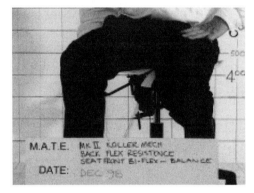

Figure 7.3. Early flat plastic prototype of seat pan for Life chair. *Courtesy of Formway Design*

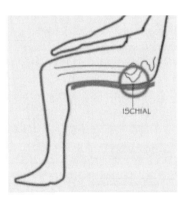 

Figure 7.4. Early prototype of Life chair concept developed to accommodate the ischial bones. *Courtesy of Formway Design*

It is sometimes possible to generate new concepts based on systematic literature and patent searches, followed by brainstorming or other group techniques to help synthesize new ideas. For example, a research contract was carried out to evaluate new technology available in the coating of coiled sheet steel under high-vacuum conditions. The better the vacuum the better the coating, and the contract called for an assessment to see if a vacuum level of 10^{-6} Torr would be possible in a full-scale operation. A massive literature and patent search indicated that the vacuum-sealing concepts used in such processes had remained much the same for many years. With the help of a single brainstorming session, several new concepts were developed. One showed particular promise, and this has now been patented as shown in Figure 7.5.

United States Patent [19]
Hales et al.

[11] Patent Number: 5,151,303
[45] Date of Patent: Sep. 29, 1992

US005151303A

[54] METHOD AND APPARATUS FOR USING EVACUATED, DETACHABLE WEB CONTAINERS WITH HIGH VACUUM TREATING MEANS

[75] Inventors: **Crispin Hales**, Winnetka; **Thomas E. Zabinski**, Orland Park, both of Ill.

[73] Assignee: **IIT Research Institute**, Chicago, Ill.

[21] Appl. No.: **564,023**

[22] Filed: **Aug. 7, 1990**

[51] Int. Cl.⁵ B05D 3/00; C23C 14/56
[52] U.S. Cl. 427/178; 427/295; 118/50; 118/718; 118/719; 118/733
[58] Field of Search 427/178, 295, 177, 294, 427/296; 118/718, 719, 733, 235, 50

[56] **References Cited**
U.S. PATENT DOCUMENTS

2,907,679 10/1959 Smith, Jr. 428/333
3,401,249 9/1968 Schleich et al. 219/69
3,990,390 11/1976 Plyshevsky et al. .... 118/719 X
4,664,062 5/1987 Kamohara et al. 118/719

FOREIGN PATENT DOCUMENTS

63-26365 2/1988 Japan .
2-82771 6/1990 Japan .

Primary Examiner—Evan Lawrence
Attorney, Agent, or Firm—Fitch, Even, Tabin & Flannery

[57] **ABSTRACT**

A coiled metal sheet is placed in a container, which may be thereafter evacuated slowly with the strip outgassing, and then the highly evacuated container is sealed with the coiled strip therein. The evacuated container may be stored until ready to be mated with a high vacuum process chamber. The container is brought to the vacuum processing chamber and is connected thereto; and the coiled strip is unwound from the container and passed through the vacuum chamber, where it is treated while under a high vacuum. The treated strip is then preferably wound in a highly evacuated take-up container.

2 Claims, 1 Drawing Sheet

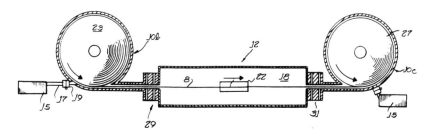

Figure 7.5. High-vacuum sealing concept

7.3 Selecting and Evaluating Concepts

Many people are creative and have lots of ideas; fewer people are good at knowing when to halt the production of new ideas, and very few have the innate ability to home in on the most promising option without some kind of simple analytical procedure. It is usually too risky to rely solely on our instincts, and there are structured procedures available for the selection and evaluation of alternative concepts. For example, a procedure for minimizing conceptual vulnerability (Pugh, 1990) is becoming well accepted and was used in selecting the concept for a new generation of space-vehicle propulsion unit.

To follow on from the guidelines for concept generation, Pahl and Beitz (1984) offer the following set of guidelines for concept selection and evaluation:

- Select suitable combinations of solution principles: use available selection techniques.
- Firm up into complete conceptual designs (concept variants): enough detail to see practicality.
- Evaluate concepts against technical criteria. Will it meet the design specification?
- Evaluate concepts against economic criteria. Is the cost low enough for viability?
- Search for weak spots. Are there detail problems that make the concept intractable?
- Select final concept(s): use available selection techniques.
- Compile cost estimates: break the concept down into individual components where possible and itemize.
- Present final concept(s) for approval: prepare carefully and present professionally.

The issues of *concept vulnerability* and *weak spots* are of particular concern to the design manager. If the wrong concept is chosen, then no amount of detail design will save it. From that moment on, the project will be in jeopardy, vulnerable to cost problems, time problems, and competition problems. Similarly, if a concept is chosen which seems excellent except for a small but intractable problem then the same may apply, and the concept must be carefully checked for weak spots using techniques such as those described in detail by Pahl and Beitz (1984, 1996).

Example: Final Seat Pan Concept for the Life Chair

The concept selected for the seat pan was a one-piece plastic membrane. Early prototype testing showed that pressure points at the leading edge of the seat pan could be reduced by cutting wave-shaped grooves at the leading edge. These slots, shown in Figure 7.6(a), created the "waterfall" effect. Similarly, the ischial bones are accommodated by the use of short slots, shown in Figure 7.6(b), cut diagonally with respect to the anterior–posterior dimension of the seat.

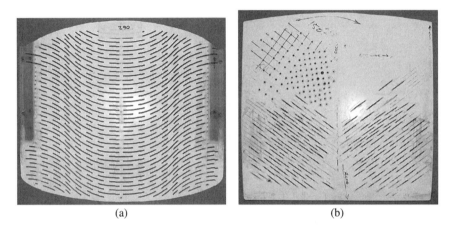

Figure 7.6. One-piece plastic membrane concept selected for Life chair. *Courtesy of Formway Design*

7.4 Estimating Costs

Budget cost estimates are needed at the concept evaluation stage, not only for use in evaluating alternatives but also to give an early warning of likely overall costs. Although there is insufficient information at the concept stage to do a precise cost estimate, very simple and quick procedures have been developed which result in surprisingly accurate preliminary estimates. One such procedure, developed from an approach used by Union Carbide engineers, is outlined below for reference purposes. It has proved very useful in practice and is formatted to allow easy updating as the final costs for each item become known. The following is the recommended procedure, with reference to the sample cost-estimate sheet for the gasifier test rig shown in Figure 7.7:

Project Name:	Gasifier Test Rig			Project Code:		Phase:	Construction		Sub-System:	Reactor Vessel	Sheet: 1 of: 9
Project Manager:	AM-A	Department:	LRS			Division:	Analytical		Date of Estimate: 22/4/83		
		Labour Hours			Purchased Materials/ Equipment	Contractor Estimate	Reserves		Confidence Level	Remarks	
Item	Work Package/Element Description	Mgmt.	Eng. / Sci.	Tech.			Specific	General			
1	Pressure Vessel Complete				18,400		2,000				
2	Safil Packed Insulation				300						
3	Safil Alumina Paper				100						
4	2 Sets Ring Elements				200		200				
5	7" I.D. Ceramic Tube				50						
6	8" I.D. Shield Tube				400						
7	12" I.D. Shield Tube				600						
8	Reactor Tube Assembly				3,000		3,000				
9	Bottom Auger				480						
10	Top Piston				50						
11	Hydraulic Actuating Cylinder				480						
12	Refractory Lining				100						
13	Specimen Shaft				500						
14	Stirrer Blades				200						
15	Central Tube				100						
16	Sampling Tube/Auger				100						
17	Shaft Bearings/Seals				200						
18	Gas Outlet Pipe Assembly				630						
19	Shaft Drive				600						
20	Bottom Outlet Pipe Assembly				200						
	Sheet Sub-Totals				26,690		2,550				

Figure 7.7. Sample cost estimate sheet

- List all sub-systems for the particular concept.
- List every known (or likely) item or component required for each sub-system.
- Prepare a series of cost estimate sheets for each sub-system.
- Prepare a title page and summary sheet to total the costs for all sub-systems.
- Estimate the complete cost of every item and assign a confidence level (high/medium/low) to each estimated cost.
- According to the confidence level, add an appropriate "reserve" cost in a separate column to cover lack of information at the time.
- Add a general item ("catch-all") to the list for each sub-system and assign a minimum 10% cost plus reserve to cover contingencies.

Note that the important thing at this stage is not the actual cost of each item so much as its inclusion in the list. Meetings, reports, and administrative work can be included, and it may be helpful to compile separate sets of sheets for different types of work effort, such as design, development, testing, and commissioning. The reserve may either be specific to a single component or spread over a number of components, and the overall preliminary cost estimate is the total of actual estimates plus all reserves. As the project proceeds, and as the final costs are obtained from suppliers or contractors, the cost estimate can be firmed up in parallel with the design itself. This is done by inserting the actual cost of an item, thereby increasing the confidence level of the estimate and hence decreasing the need for the allocated reserve. If the cost of an item is higher than originally estimated, then the higher cost is inserted and the allocated reserve reduced by an amount equal to the difference (or more if judged safe). If the cost of an item is lower than originally estimated, then the excess together with most of the reserve can be transferred to another item where there are problems, or removed altogether.

It is understandable that project sponsors and financial managers are often nervous about design cost estimates and for the design manager to uphold credibility it is extremely important to compile a working cost estimate that will not change significantly through to the end of the project. A project cost "overrun" of more than about 10% resulting from initially underestimated costs can cause tensions in the company; however, a safe (but inflated) cost estimate weakens the case for the project or product in the first place. More formal guidelines for cost estimating have been developed in Germany from years of research work on cost issues in design. These are presented by Ehrlenspiel *et al.* (1998) in German, and an English translation is in progress.

> **Example: Gasifier Test Rig**
>
> During the conceptual design the overall function of the gasifier test rig was represented diagrammatically and broken down by sub-function. Most of the sub-systems could be designed using equipment that was commercially available, but the reactor vessel assembly had to be custom designed. Five intuitive concepts evolved for this, but at the same time a systematic method for generating solutions was used. A series of eight matrices gave a large number of possible solutions, which was reduced by systematic selection and combination to four complementary matrices. Selection charts were used to decide on the most appropriate solutions, leaving three viable concepts. These matched three of the five intuitive concepts. By general agreement, the best features of each were combined into the single practicable reactor concept, shown in its original form in Figure 7.8.
>
> In theory, the output from conceptual design should be the concept that most fully satisfies the requirements of the design specification. Only those candidate concepts that satisfy every demand in the specification should pass through selection to evaluation, and then the most appropriate concept should be determined by evaluating the remaining candidates against the wishes. For the gasifier test rig this meant that any candidate concept would have to satisfy 217 demands to be selected, and those selected would have to be evaluated against 91 wishes. This presented the problem of how to deal with such a full list of requirements. In practice, the selection and evaluation procedure was based only on those requirements judged to be the most important.
>
> The structuring of the alternatives, in the form of morphological matrices and selection charts, proved to be helpful during the conceptual design of the gasifier test rig. It focused attention on the main issues during project meetings and thus simplified the decision-making process. The full evaluation procedure, involving the detailed weighting of criteria, was found to be unnecessarily complex for this particular project and highlighted the need for flexibility when applying a systematic approach.

7.5 Presenting the Final Concept

Funding the development of a concept is always a risk, and the risk will be seen as too great unless the concept clearly shows merit. This means that it must be presented to the management, customer, or both in an honest, professional, and enthusiastic fashion. Good presentations, whether formal or informal, tend to increase management enthusiasm and involvement; poor presentations do the opposite. In the case of the original "Scotts" EasyGreen® Rotary Lawn Spreader

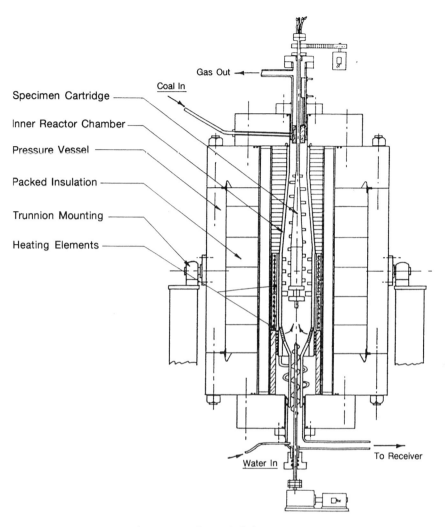

Figure 7.8. Gasifier test-rig final reactor concept

shown in Figure 7.9, a rudimentary functional model was made to demonstrate the original concept, but the marketing department immediately rejected the product idea on the basis that it would never sell. The entire project was then dropped for over a year. The design team then decided to try again, but this time they made sure that a full mock-up of the spreader as it would be sold was constructed, with the right colors and feel to the device. This time there was an enthusiastic response, resulting in full management backing for the project and a successful product for the company.

The following are a few suggestions for concept presentations:

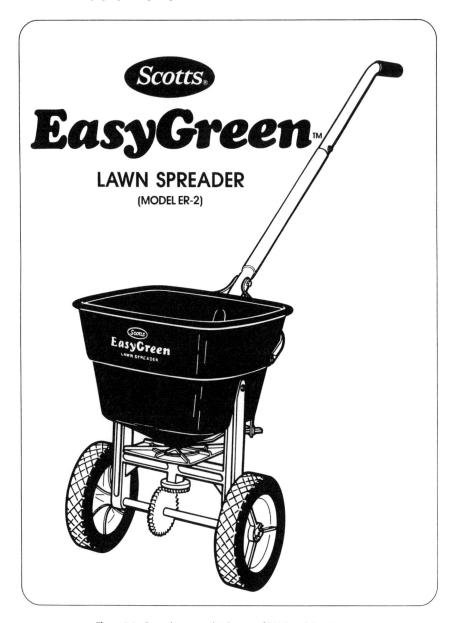

Figure 7.9. Rotary lawn spreader. *Courtesy of O.M. Scott & Sons Company*

- Carefully orchestrate and practice the presentation beforehand.
- If a computer is used, then check that the whole system is working prior to the presentation.
- Tailor the presentation to the specific audience.
- Show the concept and design work to best advantage.

- Get the main ideas across, courteously and cheerfully.
- Use simple and clear visual aids without unnecessary detail.
- Computerized presentations can be excellent, but do not let the software dominate.
- Avoid whizzing words and flashy graphics on the computer – the message gets lost.
- Models help, but they should be of high quality.
- Concisely introduce the problem, but avoid historical reviews.
- Prepare a one- or two-page summary to hand out.
- "If in doubt, leave it out."
- Do not apologize for forgotten drawings or other materials.
- Stay within the allotted time.

7.6 Conceptual Design Checklist and Work Sheet

The *Conceptual Design Checklist* shown in Figure 7.10 was developed based on the same set of headings as used for the design specification, the idea being that the "points to consider" would help the design manager to focus on the specification requirements during the conceptual design review. The design manager must ensure, from the design point of view, that a number of workable concepts have been produced and that the most promising ones have been identified and presented. The *Conceptual Design Work Sheet* provided in Figure 7.11 is intended to help the design manager review the quality of the design produced so far and to decide whether to approve it for development or require further work by the design team first. Figure 7.12 shows the completed work sheet for the seat pan concept on the Life chair.

Contributing factors attributed to the *functional requirements* were:

1. The *overall geometry* of the seat pan concept was considered to be good because it could be easily anchored to the carriage and it was unlikely to constrain the movement of the other sub-systems.
2. The *forces* could be easily determined. These forces were used to predict the maximum stresses and deflections of the seat pan using FEA. Although the stresses and deflections were found to be within acceptable limits, the team recognized that their results would need to be verified by specialist FEA engineers. This was due to the complex shape of the seat pan, with it's curved slot patterns and tapered ribs.
3. A suitable *material*, a blend of elastomeric polymers, was obtained for the seat pan. The critical properties were verified, *e.g.* by creep testing.

Safety issues evolved from the need to comply with applicable international safety standards. Underlying factors were:

1. *Operational* factors: early testing showed that the concept would meet the ultimate load tests; however, time constraints prohibited the full life testing.

CONCEPTUAL DESIGN CHECKLIST

REQUIREMENTS	CONTRIBUTING FACTORS	POINTS TO CONSIDER
FUNCTIONAL	Overall geometry Motion of parts Forces involved Energy needed Materials to be used Control system Information flow	Spatial constraints, access for assembly/operation/maintenance Practicality, accuracy, backlash, wind-up, lead and lag, smoothness Strength, stiffness, leverage, deflection, buckling, dynamics Efficiency, power source alternatives New materials, treatments, material design, compatibility, life, recycling Start-up, operation, shut-down, maintenance Input, output, storage, display
SAFETY	Operational Human Environmental	Design safety, safety standards, weak spots in design Use, misuse, outside intervention, protection, built-in safety Storage, transport, contamination, disposal
QUALITY	Quality assurance Quality control Reliability	Quality criteria, quality management, quality improvement techniques Quality measurement, quality v. cost Specified life, cost constraints, weak spots
MANUFACTURING	Production of components Purchase of components Assembly Transport	Ease of manufacture, near net shape, finish, costs Meeting specifications, transport, delivery, inspection, costs Ease of assembly, number of parts, sequencing of operation Internal transport/transfer, external transport modes, packaging
TIMING	Design schedule Development schedule Production schedule Delivery schedule	Realistic time-frame, long lead items, delay consequences Testing new technology, technological risk Tooling, long lead items Realistic time frame, field testing, commissioning
ECONOMIC	Marketing analysis Design costs Development costs Manufacturing costs Distribution costs	Adequacy of analysis, user expectations, customer expectations Historical data, design effort, team availability Test equipment, test plan, modeling, prototyping Processes involved, equipment needed, tooling Storing, packaging, transport, selling
ERGONOMIC	User needs Ergonomic design Cybernetic design	Specification requirements, types of user, different uses Conditions of use, misuse, difficulties, instructions, clarity of use Control of product in operation, runaway problems, shutdown modes
ECOLOGICAL	Material selection Working fluid selection	Recycling, disposal material, interactions, operational life Harmful effects, regulations, recycling, disposal
AESTHETIC	Customer appeal Fashion Future expectations	Field testing, surveys, national cultures, cultural differences Presentation, trade shows, timing, competition Trends, age groups, new technology
LIFE-CYCLE	Distribution Operation Maintenance Disposal	Method of distribution, advertising, promoting, national/international Life expectancy, instructions, manual, training, safety Frequency, simplicity, instructions, repair, spare parts Regulations, compatible materials, recycling, rebuilding

Figure 7.10. Conceptual design checklist

Feasible Concept: Conceptual Design   135

CONCEPTUAL DESIGN WORK SHEET

PROJECT: _____   DATE: _____

REQUIREMENTS	CONTRIBUTING FACTORS	CURRENT STATUS (Good / Marginal / Poor)	REQUIRED ACTION (Proceed / Revise / N/A)
FUNCTIONAL	Overall geometry / Motion of parts / Forces involved / Energy needed / Materials to be used / Control system / Information flow	☐☐☐☐☐	☐☐☐
SAFETY	Operational / Human / Environmental	☐☐☐☐☐	☐☐☐
QUALITY	Quality assurance / Quality control / Reliability	☐☐☐☐☐	☐☐☐
MANUFACTURING	Production of components / Purchase of components / Assembly / Transport	☐☐☐☐☐	☐☐☐
TIMING	Design schedule / Development schedule / Production schedule / Delivery schedule	☐☐☐☐☐	☐☐☐
ECONOMIC	Marketing analysis / Design costs / Development costs / Manufacturing costs / Distribution costs	☐☐☐☐☐	☐☐☐
ERGONOMIC	User needs / Ergonomic design / Cybernetic design	☐☐☐☐☐	☐☐☐
ECOLOGICAL	Material selection / Working fluid selection	☐☐☐☐☐	☐☐☐
AESTHETIC	Customer appeal / Fashion / Future expectations	☐☐☐☐☐	☐☐☐
LIFE-CYCLE	Distribution / Operation / Maintenance / Disposal	☐☐☐☐☐	☐☐☐

Figure 7.11. Conceptual design work sheet

CONCEPTUAL DESIGN WORK SHEET

PROJECT: **LIFE CHAIR** DATE: **JAN 2000**

REQUIREMENTS	CONTRIBUTING FACTORS	CURRENT STATUS (Good / Marginal / Poor)	REQUIRED ACTION (Proceed / Revise / N/A)
FUNCTIONAL	Overall geometry	Good	Proceed
	Motion of parts	Good	Proceed
	Forces involved	Marginal	Proceed
	Energy needed		N/A
	Materials to be used	Marginal	Proceed
	Control system		N/A
	Information flow		N/A
SAFETY	Operational	Marginal	Proceed
	Human	Good	Proceed
	Environmental	Marginal	Proceed
QUALITY	Quality assurance	Marginal	Proceed
	Quality control	Poor	Proceed
	Reliability	Marginal	Proceed
MANUFACTURING	Production of components	Good	Proceed
	Purchase of components	Good	Proceed
	Assembly	Good	Proceed
	Transport	Good	Proceed
TIMING	Design schedule	Marginal	Proceed
	Development schedule	Marginal	Proceed
	Production schedule	Good	Proceed
	Delivery schedule	Good	Proceed
ECONOMIC	Marketing analysis	Good	Proceed
	Design costs	Good	Proceed
	Development costs	Marginal	Proceed
	Manufacturing costs	Poor	Proceed
	Distribution costs	Marginal	Proceed
ERGONOMIC	User needs	Good	Proceed
	Ergonomic design	Good	Proceed
	Cybernetic design	Good	Proceed
ECOLOGICAL	Material selection	Good	Proceed
	Working fluid selection		N/A
AESTHETIC	Customer appeal	Marginal	Proceed
	Fashion	Marginal	Proceed
	Future expectations	Good	Proceed
LIFE-CYCLE	Distribution	Marginal	Proceed
	Operation	Marginal	Proceed
	Maintenance	Good	Proceed
	Disposal	Marginal	Proceed

Figure 7.12. Example of conceptual design work sheet

2. *Human* factors, such as misuse, had been investigated. The worst-case load was considered to be a foot or knee load on the center–middle or center–front of the seat pan. The team had shown that the seat had sufficient strength to resist this worst-case scenario.
3. *Environmental* safety was considered to have a strong influence on the safety of the seat pan.

Factors contributing to *quality* included:

1. *Quality assurance*. It was perceived that the seat pan could be made to the required quality; however, the team had not established the criteria needed to ensure that the seat pan was produced to a consistently high standard.
2. The implementation of *quality control* had not been taken into account, *e.g.* the team did not know how the seat dimensions might be measured. Although the quality of raw materials was known to be of a good standard, the cost was high and the tradeoff between quality and cost was unresolved.
3. *Reliability* had been quantified in the brief by defining a minimum working life (10-year warranty); however, weak spots needed to be verified and remedied, *e.g.* there was a concern about the shear strength of the join between the main rib along the back of the seat pan and the membrane.

Manufacturing was to be implemented on a mass-production scale. Contributing factors were:

1. The team was confident about the *production of components* because the concept had been made using the final manufacturing process, namely injection molding using a soft (aluminum) tool. A preliminary mold-flow analysis had been completed and the prototype had a good surface finish.
2. The status of the *purchase of components* factor was good because the materials were available at a competitive cost and with ISO-accredited suppliers.
3. *Assembly* was positive because the seat pan could be made in one part and it was perceived that it could be assembled into the carriage mechanism using a tool-less operation.
4. *Transport* was considered good; the seat pan has a low height profile and was likely to be efficient in terms of packing space. Production-line handling, pallet sizes, and packing issues, although considered, had not been resolved. It was perceived that these could be handled during the embodiment phase.

Factors contributing to the *timing* requirements were:

1. The *design schedule*; although development of the seat concept had taken longer than anticipated, this was consistent with the development of the other sub-system designs. The team thought that the design schedule was realistic and the consequence of delays in further development, though undesirable, could be negotiated internally with management. The main threat would be a competitor introducing a similar product into the market.

2. The level of technical risk was considered to be high; however, this was offset in the *development schedule*, which allowed time for building prototypes to prove concepts.
3. Production tooling, the testing of materials for the seat pan, and the relationships with potential suppliers evolved in parallel with the concept development. This process would need to continue as the concept evolved; the *production schedule* allowed time for this.
4. The *delivery schedule*, according to the project plan, was tight; however, the team knew that they would be able to negotiate for extra time to ensure the required quality.

Drivers for *economic* requirements were that the product be cost competitive and value for money. Contributing factors were:

1. A thorough initial *marketing analysis* was reflected in a very comprehensive design brief. Customers were involved in preliminary trialing of the seat pan concept and their needs and expectations were identified and formulated in terms of clear engineering requirements.
2. Historical data from previous projects was not relevant because of the increased level of novelty; this made the prediction of *design costs* difficult. The team did, however, have the complete support of management and a realistic budget for resources.
3. Refining the concept further required extensive prototype testing. It was thought that further refinement would involve modifying the ribbing underneath the seat pan to "fine-tune" the flexural characteristics of the membrane. This could be achieved using prototypes from the existing soft injection-molding tool and hence reduce *development costs*. If a further tool was required then this could easily be purchased within the budget.
4. *Manufacturing and distribution costs* were considered prohibitive. Although the company had a manufacturing system and distribution network that could meet the needs of the Australasian market, they did not have the resources available to create the manufacturing facilities and distribution network for sales to European and US markets. For this they planned to find a business partner.

The prerequisite for *ergonomics* was that the chair should match or better its competitors. Contributing factors were:

1. The concept scored well in terms of meeting the *user needs*. The seat pan had a soft front edge, resulting in a significant reduction in pressure behind the knee when compared with competitor products. The pan was also softer around the ischial bones. Further quantitative testing would be required to measure the actual pressure distribution.
2. *Ergonomic design* requirements were satisfied. The seat pan allowed the user to sit in a variety of postures corresponding to a wide range of tasks and activities. When working at a desk the user could be sitting "perched" on the front of the seat, sitting upright with back support, or with their legs tucked

under their bottom. The seat pan could support the user in all postures and it promoted movement.
3. *Cybernetic design* features were excellent. This was attributed to the flexibility of the seat pan being controlled by its geometry, hence requiring no adjustment by the user. Preliminary ideas for the seat support structure indicated that the flexibility of the seat could be further enhanced by the carriage design. A larger user required a stiffer seat than a smaller user. Because a large user would adjust the seat depth at a different setting from a small user, this depth adjustment function could be incorporated to provide the seat pan with increased stiffness when in the "out position" (large user) and less stiffness when in the "in position" (smaller user). The carriage could also be designed to include a fail-safe mode, where the seat would bottom-out under very high loads.

Ecological requirements were governed by materials selection, which was considered to be marginal to good. The plastic was recyclable; however, the blend of two materials reduced its value. Introducing a product stewardship programme, where the company retrieved chairs at the end of their working life, could be introduced to accommodate this effect.

One of the primary drivers for this project was an "inspirational and uplifting *aesthetic.*" Factors influencing aesthetic requirements were:

1. *Customer appeal*: the seat pan concept was good in terms of customer appeal. The slotted patterns and tapered rib gave the appearance of a high-tech/high-performance design and, therefore, high perceived value.
2. *Fashion* was considered to have a significant influence on the product's success. The seat pan could be easily dressed up by adding a seat cover to change the aesthetic to meet current trends. This also made the chair compatible with both domestic and commercial environments.
3. *Future expectations* were good for the seat pan. Trends in other office furniture products were moving to a light, minimalist aesthetic. The designers felt that competitor products were not meeting this need. The minimalist aesthetic also fitted environmental trends.

Good environmental performance over the product *life-cycle* was considered critical, along with the products ability to meet the needs of the future office environment. Other contributing factors included:

1. *Distribution*: this was yet to be resolved. The company could supply the local market; however, an international partner with existing distribution networks would be needed to distribute the product in Europe and the USA.
2. *Operational* factors were considered good, because the seat pan was expected to have a long working life. Its surface was durable and the plastic had good fatigue properties.
3. *Maintenance* was also good. In the event of a failure, this part could be removed and replaced by the end user without the need for tools.

4. *Disposal* factors had a good influence on life-cycle requirements. The seat pan could be disassembled from the chair and was uncontaminated by other materials (*i.e.* no fastenings or other materials were present).

The completed work sheet shows a good level of confidence in the areas of manufacturing, economics, ergonomics, and life-cycle usage. By this time, then, the design team could expect enthusiastic support from management, marketing, and manufacturing, which would encourage them to improve the concept while investigating the means for producing and distributing this product on a mass-production scale.

7.7 Tips for Management

- Understand divergent and convergent thinking.
- Abstract the problem to get to the real issues.
- Make sure many different ideas are generated, never just one or two.
- Use available methods and techniques for systematic selection and evaluation.
- Make sure the right concept is chosen.
- Search for weak spots in the final concepts and understand the implications.
- Produce an itemized budget cost estimate, reflecting the current degree of confidence.
- Present the finally chosen concept in a formal and professional manner.
- Consider patenting novel concepts.
- Use the checklist and work sheet to review the design work prior to approval.

Chapter 8
Developed Concept: Embodiment Design

8.1 Abstract Concept to Developed Design
8.2 Overall Guidelines for Embodiment Design
8.3 Specific Guidelines for Embodiment Design
8.4 General Guidelines for Embodiment Design
8.5 Embodiment Design Checklist and Work Sheet
8.6 Tips for Management

8.1 Abstract Concept to Developed Design

There is a huge jump from the visualization of an abstract concept to the manufacturing drawings from which a safe and reliable product can be made. If this jump is made without sufficient thought and without appropriate development of the ideas, then design failure is almost certain. The jump may be made in a number of ways. Commonly, a prototype is built and put through a rigorous test program, which is a costly and time-consuming exercise. Technically, it is well worthwhile when mass-production is contemplated, but the tendency towards shorter and shorter development times may make it economically unacceptable. New products are now often introduced right on the production line by employing an *incremental design approach*. This involves systematically introducing new technology or redesigned components on existing products to test and prove them in practice, rather than taking the more risky approach of introducing everything at once in a completely new product. Not only has much of the new product then been through field testing, but the overall development time is dramatically decreased and it is easier to identify the cause of any problems that may arise. If the project is a "one-off," or if very small production numbers are involved, then prototype testing of any kind may be out of the question. It may be possible to simulate the final performance on a computer, and it may be possible to gain sufficient confidence in the overall design by testing certain components only. Whatever course of action is appropriate, a lot of design effort is needed between the approval of a design concept and the final detailing of parts for manufacture. For the design manager, it is helpful to divide this large part of the design effort into more manageable phases, and for the purposes of this book two phases will be used, termed *embodiment design* and *detail design*.

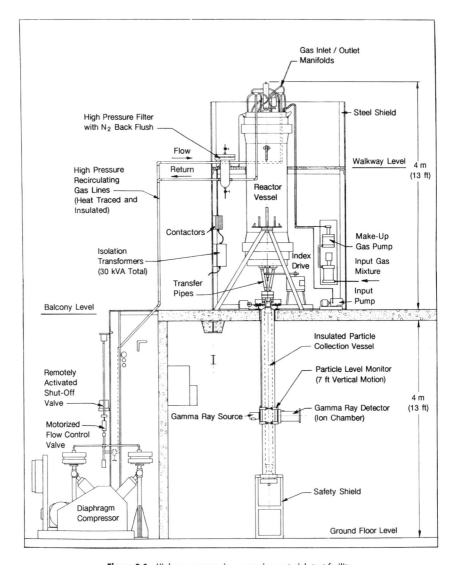

Figure 8.1. High-pressure erosion–corrosion materials test facility

To illustrate what is meant by embodiment design, as distinct from detail design, let us consider the design of the main reactor vessel for an erosion–corrosion test facility (Hales *et al.*, 1981), shown diagrammatically in Figure 8.1.

This test facility was for evaluating the erosion–corrosion performance of superalloys at high temperatures (980°C) and pressures (100 bar) in a simulated coal-gasification environment. The principle of operation was to entrain erodent in the recirculating high-temperature corrosive gas stream and impinge this through 12 jet tubes onto test coupons (specimens) of the candidate alloy materials mounted at 45° to the gas stream. Eleven materials could be tested at the same time, and to ensure that each coupon was subjected to the same impingement the coupons were mounted on a circular holder, periodically indexed 30°. The twelfth position was used for control and to enable the sampling of the combined flow of gas and erodent from each jet in turn. The concept for the reactor vessel originated from the research group in Chicago contracted to carry out the metallurgical test program, and the final concept was drawn up as shown in Figure 8.2.

As a conceptual drawing it is good. It illustrates the principle of using an outer vessel to contain the high-pressure gases, with an inner vessel pressure-balanced across its wall to accept the very high temperature without rupture. It also shows the idea of mounting the test specimens at 45° on a platform indexed beneath a series of jets through which the gas and erodent flows. The concept was considered innovative and the project sponsors were sufficiently confident that it would work to approve construction of two reactor vessels.

The conceptual drawing shown in Figure 8.2 was given to a company draftsman who simply produced detail drawings of every component as itemized on the conceptual drawing, without further design. The internal parts were manufactured directly from these drawings, mostly from expensive high-grade stainless steels and superalloys. Meanwhile, the drawings of the pressure-vessel components were sent to the manufacturer in Texas, who, in turn, produced a set of manufacturing drawings. The manufacturing drawings were officially approved by a professional engineer as meeting the applicable ASME Boiler and Pressure Vessel Code for Unfired Pressure Vessels (Section VIII, Division 1) under partial stamp requirements. The only change made was in the use of studs and nuts for the top and bottom closures (Code requirement) instead of the "Hold Down Bolts" shown in Figure 8.2. The two pressure vessels, weighing about four tons each, were eventually manufactured and delivered, the cost being approximately US$80 000 for the pair. Without knowing any more details about the system, a series of interesting features in Figure 8.2 will be apparent to the reader, such as:

- The vessel is shown balanced on a floor-mounted scissors jack (to represent vertical shaft adjustment).
- There are no supporting brackets or mounts for the vessel.
- The volume of the top cover plate (over the hole) is less than the volume of material removed from the top blind hub.
- The only seal shown for the rotating specimen carrier shaft is a single elastomeric O-ring.
- This seal has to prevent corrosive high-temperature gas inside the reactor from escaping.
- The only bearing shown for the rotating specimen carrier shaft is a graphite bushing.

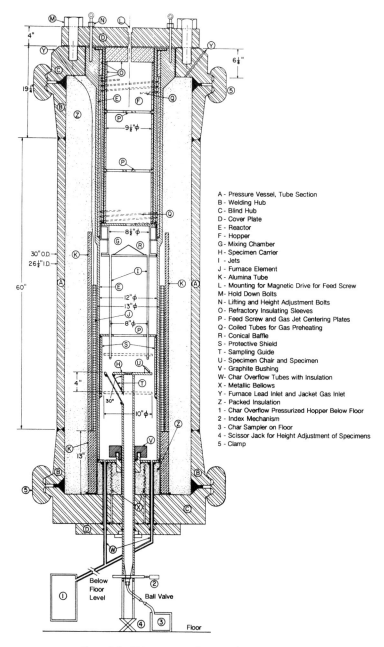

Figure 8.2. High-pressure erosion–corrosion reactor concept

- The alignment of the jets is by means of a plate located by the wall of the reactor tube.
- The vertical and horizontal alignment of the specimen carrier shaft depends on the differential movement between the reactor tube, the pressure vessel, and the ground.

It was at this stage, when all the internal parts had been fabricated and the vessels were due for delivery, that the company recruited a design engineer (Hales) to review the design and to complete the system. It had already been realized that the vessels had no physical means of support, and that the ASME Code forbade any further welding once the vessels had been heat-treated! This oversight had been dealt with quickly by making four one-inch thick flat plate arms for each vessel and shipping them to Texas for welding to the outside of each vessel shell prior to heat treatment. Each arm had a one-inch diameter hole drilled through it prior to welding and, despite there being no specified tolerances, these four holes were supposed to correspond in location precisely with four one-inch diameter holes already drilled in the structural steel frame in Chicago! At that time it had been planned to use four one-inch diameter bolts to locate and support each vessel in its frame, the bolts, therefore, having to pass through aligned holes of the same diameter. A review of the complete design revealed a series of major deficiencies. The pressure vessels were inadequately designed for the operating conditions, the flat plate brackets were inadequate for lateral support, internal and external assembly would be physically impossible, and, even if the components could be assembled, the seals and bearings could not function as drawn. Almost all the funding was gone, the time was gone, detail drawings had been prepared for each "conceptual part," and all these parts had been machined! In addition, during a review of the certification papers for the vessels, it was discovered that the two vessels had been improperly heat-treated. The records of the actual heat treatment did not match the heat treatment procedure specified. This meant that the two vessels had to be shipped back to Texas at the manufacturer's expense to be heat-treated again. As it happened, this manufacturing error was fortuitous, as it provided the opportunity to correct some of the major design problems at little cost to the project. By designing new substantial and stable bottom-support brackets for welding to the shell prior to reheat treatment, a reference datum for the entire vessel and reactor assembly was established. The vessel closures were redesigned in conformance with special flat plate closure requirements of the ASME Boiler and Pressure Vessel Code. Each of the circular stacked flat plates was used as the base plate for a cartridge type of sub-assembly, thus converting the entire internal arrangement into a series of independent modules with a common reference datum. The components within each cartridge sub-assembly were designed to expand or contract relative to their own base plate, and mechanical interaction between modules was minimized. At the same time, the component details were determined precisely to provide maximum dimensional stability for jet/sample alignment with minimum transfer of heat away from the reactor hot zone. In this way it was possible to maintain an excellent alignment between each jet and the samples indexing beneath them, despite the near white-heat operating temperatures. Many of the parts already made were reused as semi-finished raw material for the redesigned components. For example, the original top cover plate was remachined to form the new bottom cap, housing the main sliding seals and bearing cartridge for the central hollow indexing shaft. The final embodiment design, as developed from the original concept, is shown in Figure 8.3. Relative to the reference datum at the lower support of the vessel, the external shell expands upwards towards the top closure system and downwards towards the bottom closure system. At the same time, inside the vessel, the subsystem components in those cartridges hanging down from their bolted base plates at the top expand downwards, while those supported on their bolted base plates at the bottom expand upwards. In order to accommodate this complex series of differential expansions, and to enable removal of the lower cartridges without affecting the upper ones, the indexing shaft within the inner reactor was driven from a separate lower shaft by means of a sliding dog-type coupling arrangement. This is shown in more detail, together with the sliding seals, in Figure 8.4.

The embodiment design features, as distinct from the conceptual design features, include:

- Stable bottom bracket of vessel provides datum for expansion/contraction calculations.
- Erodent hopper, mixer, and jet sub-assembly form a cartridge unit hung from the top cap.
- Jet alignment controlled from dowelled top cap of vessel by means of rigid cartridge unit.
- Ceramic jet tubes used, supported and aligned by a highly rigid hollow central column.
- Each jet may be aligned and checked as a sub-assembly task, prior to vessel assembly.
- Reactor assembly designed as a cartridge, aligned and fitted from the top of the vessel.
- Reactor held in alignment with respect to vessel at bottom by means of a sliding bushing.
- Reactor wall thickness optimized for adequate strength with minimum heat transfer.
- Modular assembly of specimen holder with dust shield, bearings, and seals within reactor.
- Cartridge sub-assembly of high-pressure nitrogen seals and lower drive shaft in bottom cap.
- Modular sub-assembly of sliding seals and drive shaft in bottom cap.
- Isolation of high-temperature reactive gas within reactor by means of seals operating only at low differential pressure between reactor and vessel.

Developed Concept: Embodiment Design 147

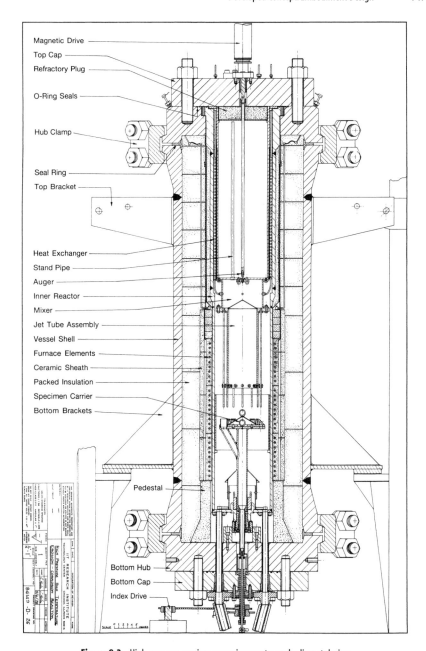

Figure 8.3. High-pressure erosion–corrosion reactor embodiment design

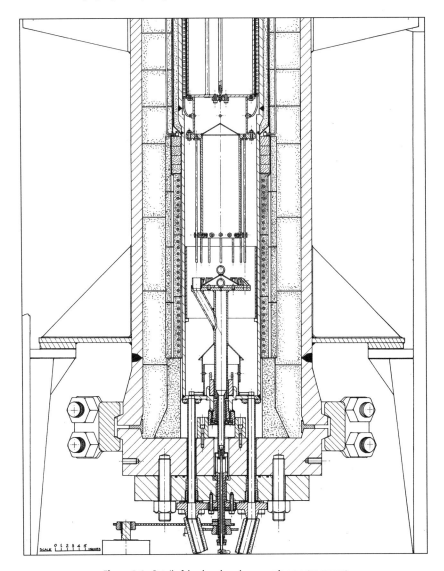

Figure 8.4. Detail of developed erosion–corrosion reactor concept

- Leak detection system between double O-ring seals at top of vessel, with the facility for pressurizing with nitrogen between seals in the event of a leak during operation.

It will be noted that such features are concerned not with the detailed dimensioning of individual components for manufacture, but with the establishing of reference datums, the logic of component and sub-assembly alignments, the control of force paths and heat transfer paths, the grouping of components into sub-assemblies for ease of assembly and maintenance, and with the use of cartridge-type sub-assemblies for practical operation.

Based on this work, the detail design of every component was then completed by the author and all parts were manufactured according to the replacement design drawings. The reactor vessels were assembled, installed together with the complex gas pumping and control systems, and were commissioned into long-term service on tests of up to 1000 h duration. Although the system performed successfully in the end, the 1 year time delay and the enormous cost associated with the redesign severely hampered the testing program. The problems could have been avoided had specific embodiment guidelines been applied prior to detailing the components for manufacture. This example typifies the problems that arise when a *conceptual* design goes straight to *detail* design, missing out the *embodiment* design phase. The embodiment part of the design process takes a lot of time; it is critical to the success of the design, and the design manager needs to be confident that the embodiment design issues have been adequately addressed before the expensive business of detailing every component for manufacture is initiated. In practice, there should be considerable overlap between embodiment design and detail design, but in the mind of the design manager there must be the clarity of thought that will ensure that embodiment design issues are resolved before detail design of components is carried out.

In order to help the design manager to make sure that embodiment design issues are adequately addressed, some useful guidelines have been collected together from various sources. These are summarized in the following sections. The guidelines have been divided into three groups: the first two are based on the approach of Pahl and Beitz (1984, 1996) and the third is added from one of our own personal experiences. The guidelines are presented here in summary form only, providing the basis for a checklist and work sheet to help the design manager in assessing the status and quality of embodiment design during the course of a project. For a more detailed description of the guidelines and their application in practice, the reader is referred to the Bibliography.

8.2 Overall Guidelines for Embodiment Design

Sufficient design "theory" now exists to provide a reasonably systematic approach to the development of mechanical design concepts, and certainly to provide guidelines and a simple checklist to help the design manager evaluate the quality of a design before committing it to detailing and manufacture. During embodiment design, the aim is to resolve overall geometrical, dynamic, and safety issues, and to develop more complete layouts of the concept by consideration of each assembly, sub-assembly, and component in turn. Questions need to be answered, such as:

- Will it work?
- Is it safe?
- What function does it serve?
- Will it be made from scratch, bought in, or made from semi-finished material?
- How does it fit in with the rest of the design?
- What development will be required?
- How long will it last?
- How might it fail in practice?

8.2.1 Clarity

Clarity of function, form, and layout should be considered, and used to advantage. It should be quite clear what purpose each component or sub-assembly serves, how it is to be manufactured, and how it fits together with all the other parts. For example, in Figure 8.4 it is clear that the bottom bushing of the inner reactor vessel fits into the bushing mounted inside the bottom hub and that this restrains the inner reactor horizontally, keeping it in vertical alignment while allowing vertical expansion and contraction of the reactor tube as it is heated and cooled within the pressure vessel. In the conceptual design, shown in Figure 8.2, it is not clear how the reactor will move relative to the pressure vessel.

8.2.2 Simplicity

The simpler that the arrangement and the shapes used are, the lower the cost of manufacture and the better the overall design is likely to be. The aim is to use the minimum number of components with the simplest possible shapes. Although clarity and simplicity are closely related, they are not the same and must be considered separately. A product may be extremely simple, such as a single-component hand tool, yet its function may be obscure. Simplicity is an aid to clarity, and in achieving clarity it is likely that a design will become simpler, so there is a "chicken-and-egg" effect with these two factors. The use of a cartridge-type sub-assembly for the reactor in Figure 8.3 is an example of operational simplicity. To remove the specimens from the complicated reactor, the operators simply undo the nuts holding down the top cap and withdraw the cartridge, including hopper, mixing chamber, and jet tube sub-assembly. This gives immediate access to the specimens *without affecting the alignment of the jet tubes or the geometrical relationship between the jet tubes and the specimens*. Thus, the alignment and location issues are no longer problems from the operator's point of view.

8.2.3 Safety

Safety in design has become increasingly important (Hunter, 1992). Commercial pressures often demand that products are used to the limits of (or beyond) their designed capability, yet if there is a failure the the design engineer becomes liable for damages. It is over matters of safety and product liability that the advantage of having used a systematic and well-documented approach to design becomes obvious. Clarity and simplicity automatically contribute towards safety in design, but there is also an established safety hierarchy for design, as described generally by Pahl and Beitz (1984, 1996) and in more detail, for example, by Barnett and Switalski (1988). The safety hierarchy is aimed specifically at minimizing the harmful effects of component, system, or operational failures as follows:

- First priority   – eliminate the hazard and/or risk (direct approach).
- Second priority   – apply protective systems (indirect approach).
- Third priority   – provide warnings.
- Fourth priority   – provide for training and instruction.
- Fifth priority   – prescribe personal protection.

Safety, as used here, means more than human safety. It includes situations where the failure of one component may adversely affect the performance of another and lead to operational or reliability problems, whether or not human safety is involved. For example, although the failure of the Space Shuttle Challenger may be regarded as a human safety problem because humans were on board, there were a series of design weaknesses that would be regarded as safety or reliability issues in design even if the vehicle had been flying with no crew. One of these was the fact that once the hot, pressurized gases within the solid rocket booster had reached the aft field joint elastomeric O-ring seal, the heat damaged the seal, thereby letting more gas flow, thereby damaging the seal more, thereby creating a bigger leak, and so on. In embodiment design this is regarded as an *unstable* design, for which there are specific remedial guidelines. As the hazard had not been eliminated, the second priority would have been to design some kind of protective system. In fact, a second, back-up O-ring seal was intended to serve this purpose; but, as it has the same performance characteristics as the first one, at best it could provide only a time delay before failure. One alternative approach might have been to create a secondary seal by using the fact that steel expands on heating. As the temperature increased, the secondary seal would automatically take over from the damaged O-ring seal and the sealing performance of the arrangement would actually increase with temperature.

Each level of the safety hierarchy is briefly considered in turn to provide more detailed information as background to the checklist and work sheet at the end of this chapter. In keeping with the rest of the book, the aim is not to show how to apply the guidelines in design practice, but to help the design manager make sure that all the important aspects have been addressed before the design is approved.

8.2.4 Direct Approaches to Safety and Reliability in Design

8.2.4.1 Safe-life Design

Components and their connections are designed to survive their predicted working life without failure. By this is meant without fracture or critical deformation. If failure occurs, then an accident is possible, such as with a helicopter rotor blade or a bicycle brake cable. However, the likelihood of failure is reduced by:

- clearly specifying all operating conditions;
- careful attention to detail design, including application of safety factors;
- provision of overload capacity;
- regular maintenance and inspection during operation.

8.2.4.2 Fail-safe Design

After a failure, some capacity to perform critical functions should remain. Failure should be signaled (directly or indirectly) so that the system can be shut down safely. For example, wear in bearings (noise and vibration) or a blown head-gasket (water in lubricating oil). During the design work, possible failures must be foreseen and steps taken to ensure that the consequences are acceptable or can be made safe.

8.2.4.3 Redundant Design

Additional elements or systems are provided to take over the function in the event of a failure. Note that the possibility of a first failure causing a catastrophic failure by also damaging the redundant system must be avoided, such as an explosion that damages a stand-by system. Redundancy can be:

- parallel   – active or passive;
- series   – active or passive.

Examples are:

- parallel active   – aircraft with more than one engine;
- parallel passive   – stand-by generator;
- series active   – double in-line filters;
- series passive   – anti-lock braking unit.

8.2.5 Indirect Approaches to Safety and Reliability in Design

The indirect approach involves providing protective systems, *e.g.* electrical fuses; overspeed devices; safety valves; alarm systems; machine guards; and sprinkler systems. The protective system should, if possible, be self-monitoring, with the capability to detect and act on faults by itself. Safety is enhanced if the protective system includes redundancy, *e.g.* two systems working either in parallel or in series. Safety is further enhanced if the two (or more) protective systems work on different principles. This will avoid failure of each protective system due to common faults such as corrosion.

8.2.6 Warnings

This approach involves providing warnings of potential danger by means of notices, signals, or barriers. This third level of safety measure should be used only as backup. Direct or indirect safety measures should be incorporated whenever possible and in preference to warnings. In fact, warnings are rarely effective in practice (Barnett, 1998), but they have become of great importance because of the legal profession. Plaintiffs' attorneys have used the "warnings issue" to the point of excess in generating claims and negotiating settlements in product liability lawsuits (American Law Institute, 1998).

8.2.7 Training and Instruction

Many products rely on training of the operator as an integral part of design for safety, and thought must be given to this during embodiment design. The driver of a car, for example, must be trained in vehicle handling as well as in learning how to behave and cope with typical driving situations on the road. For simpler products, instruction manuals are sufficient, but great care must go into their design in order to ensure that the product is used correctly and that clear warnings are provided to cover foreseeable misuse. The legal profession, particularly in the USA, has created a system where the design engineer is likely to be faulted even in the most extreme of misuse situations, and the design engineer must be able to prove that everything possible was done to ensure the reasonable safety of product users.

8.2.8 Personal Protection

There are also situations, such as use of sandblasting equipment, where it is essential for operators to wear protective clothing and devices to ensure safe working conditions. This then becomes part of the design of the system, and the design engineer must treat it as such. It may require the services of safety experts to advise on how best to meet the applicable regulations and standards, as the issues can be subtle and outside the scope of the expertise of the design engineer.

> **Example: Space Shuttle Challenger**
>
> Taken from *Analysis of an Engineering Design: The Space Shuttle Challenger* (Hales, 1989).
>
> **Embodiment Design Questions Concerning the Solid Rocket Motor Aft Field Joint:**
>
> 1. Is the design simple?
> 2. Is there clarity of function?
> 3. Is there clarity of form?
> 4. Is it safe – safe-life design?
>   – fail-safe design?
>   – redundancy built in?
>   – protection built in?
>   – warnings provided?
>
> **Evidence Based on the Report of the Presidential Commission:**
>
> 1. Joint has complex loadings, geometry, sealing, and thermal conditions.
> 2. Load paths for forces not clear and operation of seals not clear.
> 3. Confusion over tolerances and joint gap.
> 4. Not safe-life: O-rings operating outside recommended conditions.
>   Not fail-safe: failure of primary O-ring allows hot gases access to secondary O-ring.
>   No redundancy: joint redesignated Criticality 1.
>   No safety protection from gas channeling.
>   No safety protection for strut connecting rocket to external tank.
>   No safety warning of seal/joint failure.

8.3 Specific Guidelines for Embodiment Design

The above overall guidelines help to provide a structured approach to the development of any design concept. Associated with the overall guidelines have been identified a number of detail or specific guidelines (sometimes termed "principles") to help with particular aspects, as described by Pahl and Beitz (1984, 1996):

- Force transmission
- Division of tasks
- Self-help
- Stability

Each of these is introduced briefly below, together with some examples.

8.3.1 Force Transmission

8.3.1.1 Flowlines of Force

The function of many components is to transmit forces (and moments) from one point to another. A simple analogy to help visualize force transmission is in flowlines of force as illustrated Figure 8.5, adapted from Wallace (1984). The external loads applied to a structure are balanced at every section by internal forces and moments. At critical sections the stresses can be calculated and compared with the material strength. A desirable aim is *uniform strength*, even if it cannot be achieved. Guidelines for uniform strength include the following:

- Avoid abrupt changes of cross-section.
- Avoid sharp changes in direction.
- Avoid changes in flowline density.

8.3.1.2 Force Transmission Paths

Sometimes the minimum deformation is required of a product in service, whereas at other times large elastic deformations are required. The base of a machine tool might be an example of the first, and an isolation spring is an example of the second. Guidelines for minimum deformation (rigid components) include:

- Use shortest possible force transmission path.
- Use most direct force transmission path.
- Work with axial or shear forces.
- Favor symmetrical layouts.

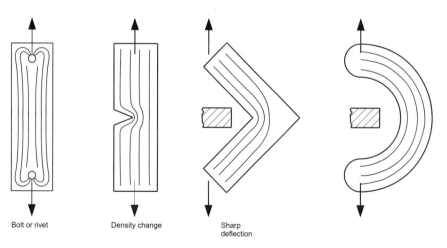

Figure 8.5. Visualizing flowlines of force (adapted from Wallace, 1984)

Guidelines for large elastic deformations (flexible components) include:

- Use a longer force transmission path.
- Use an indirect force transmission path.
- Work with bending or torsional moments.

8.3.1.3 Matched Deformations

Mismatched deformations between related components can lead to uneven stress distributions and unwanted stress concentrations. For example, differential torsional wind-up in a gantry crane drive axle when the drive is from one end of the shaft leads to jerky longitudinal crane motion (Pahl and Beitz, 1984). Thus, an important guideline for matching deformation is:

- Design interacting components so that, if possible, they deform in the same sense and by the same amount under load.

For example, the crane shaft could be driven from the center, or the torsional stiffness of the longer driveshaft could be increased by a larger shaft diameter.

8.3.1.4 Balanced Forces

The main forces produced within a mechanical system as part of its function often give rise to secondary, associated forces that must be accepted within the system even though they do not serve the function directly. Guidelines for balancing out associated forces include:

- Balance out associated forces close to their point of origin whenever possible.
- Use additional force-balancing elements for medium loads.
- Use symmetrical layouts for high loads.

8.3.2 Division of Tasks

A component may be designed to serve just one specific task, such as a dowel, or a component may be designed to perform several tasks, such as an engine drive pulley. Several identical components may be required for high-load situations, such as multiple Vee-belts. This gives rise to a series of helpful guidelines, together with constraints, as follows:

The main guideline for combining tasks is:

- Assign several tasks to a component for economy in space, weight, number of parts or cost, *but note*
    - it may compromise the performance of individual functions;
    - it may require design and analysis of more complicated shapes;
    - it may be more expensive to replace if there is a fault.

Guidelines for division of tasks include:

- Assign a specific component to a specific task for optimization or if the task is critical.
- Use several identical components to cope with loads or size too great for one, *but note*
  - increased space and weight involved;
  - more parts and connections used;
  - there is difficulty in getting identical parts each to carry the same load.

8.3.3 Self-help

The idea of self-help is to improve the performance of a function by the way in which components interact with each other. It can provide a greater effect, a reduced effect, or greater safety (in overload conditions), depending on the circumstances and what is required:

Overall effect = Initial effect + Supplementary effect

8.3.3.1 Self-reinforcing

The required effect increases with increasing need for the effect, such as better sealing of O-rings as pressure increases. Leading shoe drum brakes are self-reinforcing, in that the braking force increases rapidly with increasing pedal pressure (tendency to "grab"). With disk brakes there is also a self-reinforcing effect, but through a different mechanism. The braking force increases as the brake disk expands with heat.

8.3.3.2 Self-damaging or Self-balancing

In this case the supplementary effect reduces the initial effect. For example, with a trailing shoe drum brake an increased force is required to maintain braking effect as the brake heats up. This can be used to exercise control over the grabbing tendencies of a leading shoe brake and a stable system results when a leading shoe is combined with a trailing shoe design.

8.3.3.3 Self-protecting

"Self-protecting" means that components should be designed to survive in the event of an overload, unless intentionally used as weak links. By providing an additional force transmission path it is possible to alter the flowlines of force after, for example, a given elastic deformation, so that the load is still carried without component damage. This is termed a self-protecting solution, and an example would be the bump stop on a car suspension spring.

8.3.3.4 Summary

Guidelines for self-help include:

- For self-reinforcing solutions use primary or associated forces acting in the same sense as other main forces.
- For self-balancing solutions use associated forces acting in the opposite sense to primary forces.
- For self-protecting solutions change the force depending on elastic deformation *and note* that self-damaging effects can easily be produced which are usually, but not always, detrimental.

8.3.4 Stability

The stability of a design can be considered on a number of different levels, but in this particular context the important thing is whether the designed system should, and if so will, recover appropriately from a disturbance. For example, the physical capsize of the *Herald of Free Enterprise* ferry in 1987 was caused by a set of cumulative disturbing factors, including turning forces from steering, lateral water slosh, lateral sliding of vehicles, and inversion of the center of gravity with the center of buoyancy. The effects of the disturbances increased with angle of roll to the point where there could be no recovery and the ferry turned onto its side. At the time there were no design features that opposed or canceled the effect of these additive disturbances, so the design under these circumstances was inherently unstable. If the bow doors had not let water in, if the deck design had prevented water slosh, if the vehicles had been tied down, and if the trucks had been less heavily laden, then the ferry may have been able to recover from the effects of the turning forces from steering.

The main guideline for stability is:

- Consider the effects of abnormal disturbances and ensure that these effects will be reduced or canceled out.

Sometimes planned *instability* is useful, such as the over-center toggle action of an electric light switch to ensure that the switched is either on or off and cannot stay in the middle neutral position. Another example is the "poppet" type of safety valve, which snaps fully open when a limiting pressure is reached.

The main guideline for use of planned instability is:

- Introduce self-reinforcing effects when a selected physical quantity reaches a limiting value.

Example: Space Shuttle Challenger

Taken from *Analysis of an Engineering Design: The Space Shuttle Challenger* (Hales, 1989).

Further Embodiment Design Questions Concerning the Solid Rocket Motor Aft Field Joint:

1. How good are the force transmission paths – flowlines?
   – deformations?
   – secondary forces?
2. Is the division of tasks appropriate (separated, combined, or divided)?
3. Is self-help used – self-reinforcing?
   – self-balancing?
   – self-protecting?
   – what about self-damaging?
4. Is the design stable?

Evidence Based on the Report of the Presidential Commission:

1. Force flowlines concentrate around pins – poor load distribution. Joint rotates, affecting seal performance. Ice and putty create detrimental secondary forces.
2. Critical tasks of load-carrying and sealing are combined instead of being separated.
3. Self-reinforcing used, but minor gas leaks create a self-damaging situation.
4. Design unstable – joint gap leads to seal leak
   – joint rotation leads to bigger leak
   – putty allows gas channeling
   – gas channeling leads to burnt seals
   – burnt seals lead to bigger leak
   – bigger leak leads to high temperatures at joint
   – high temperatures lead to hole in casing
   – hole in casing leads to strut failure
   – strut failure leads to catastrophic failure.

8.4 General Guidelines for Embodiment Design

8.4.1 Use of Calculations

Calculations are an essential part of engineering design work, and many are usually required. They should be carried out and recorded in a careful manner for later reference, and computer printouts should be carefully annotated and documented for easy understanding by other engineers. One of the difficulties with the analysis of the tri-axis transfer press design mentioned in earlier chapters was the lack of recorded calculations. All that existed were some preliminary hand calculations and some undocumented computer printouts of later calculations. The in-house program used for the computer calculations had been modified and adapted to such an extent that the original programmer was no longer able to provide the necessary documentation retroactively.

A few guidelines for use of calculations are as follows:

- Approximate calculations should be used throughout the embodiment design phase to determine effects and consequences (gain insight).
- More accurate calculations (and repeat of first-order calculations) should be carried out as the layout and form design is firmed up (see French, 1999).
- Calculations should be completed on standard format sheets, which include the date, project nature, title of calculation, assumptions made, and symbols used. Origin of equations, etc. should be referenced in a right-hand margin.
- Calculations should be indexed, checked (by another person), and retained with the project file.

Example: Seat Pan for the "Life" Office Chair

The front edge of the seat was designed to flex in order to reduce pressure points behind the knees. The seat pan also included localized "soft" spots to accommodate the ischial bones, reducing both pressure points and the tendency to slip forward in the seat. These concepts were proven using early prototypes, which showed that flexibility could be achieved by making slotted holes in the seat. Further development of this concept required and investigation into the effect of various hole patterns on front-edge deflection and ischial bone accommodation. Prototyping was considered too expensive and time consuming. The team used Finite Element Analysis (FEA) to predict deflections and stress levels for numerous different hole patterns. Although they were experienced in the use of FEA, the team employed external consultants to check the analysis using independent FEA methods before final prototypes were manufactured. These prototypes were built using final materials and production processes. Figure 8.6 shows a deflection plot for one of the final seat pan designs as produced by an external consultant. A color reproduction of this plot is provided. See also color plate provided.

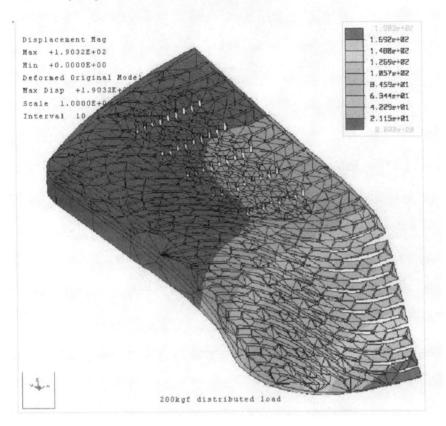

Figure 8.6. FEA plot for the Life chair seat pan. Courtesy of Formway Design. See plate section for color version.

8.4.2 Materials Selection

Selecting materials in design was once a relatively simple matter, but with the rapidly increasing range of new materials and complex alloys the task of selection has become difficult and demanding. Improved materials can lead to better products, easier manufacture, longer service life, and lower cost. It used to be that a large number of brochures, handbooks, catalogs, directories, and materials specialists had to be consulted to ensure that the most appropriate material was found. However, during the past decade there have been enormous advances in the materials information available from databases and the Internet, with built-in procedures to assist the design engineer in the selection process. For example, the Cambridge Engineering Selector (Cebon, 2003) now covers manufacturing process information as well as cost, availability, form, and physical data. Essentially, it is becoming possible to design a material to suit a specific application as well as to select a material most appropriate to a particular design. These developments should help to overcome:

- a tendency to avoid use of new or alternative materials;
- inappropriate use of materials;
- waste of materials resources.

However, it should be stressed that great care is needed when selecting a material that may not have been tested sufficiently under real-life conditions for a particular application. For example, if a plastic type of material is selected for an application where traditionally metals have been used, then secondary problems, such as inadequate stiffness, degradation from exposure to ultraviolet light, and a greater likelihood of failure in the event of a temperature excursion, may offset advantages such as lower manufacturing cost and better corrosion resistance. For simple materials selection, the following is a suggested guideline:

- Determine basic form of component.
- Determine form, availability, and cost of several candidate materials (raw or semi-finished).
- Use computer-based selection procedure or Internet product information search, as appropriate.
- Match form and manufacturing process of available material against form design of component.
- Select the most appropriate materials and check physical properties against basic functional requirements.
- Iterate if necessary.
- Refer to database or data sheets for material details and make final selection.

8.4.3 Design Standards and Codes

In general, design standards encapsulate what has become accepted best practice for the design of particular types of product, and they provide at least a baseline set of criteria for the evaluation of designs produced, as discussed further in Chapter 11. Common ones are:

- Company standards;
- Industry standards;
- National standards;
- International standards.

Compliance with applicable standards is strongly recommended, *as a minimum requirement*, even for those considered to be "voluntary standards." In the event of a safety or performance problem with a product, the very first questions are likely to be:

- What standards were applicable to the product?
- Does the product comply with current standards?
- If so, where is the evidence of this?
- If not, why not?

The design manager, called in to testify regarding the quality of a product design, starts in a weak defensive position if an applicable standard existed yet it was not complied with. Not to know of the existence of such a standard is usually indefensible. Standards are often limited in scope and may contain conflicting requirements or ambiguities. In this case, an interpretation should be sought and kept in the records for future reference. Some standards have been developed based on insufficient practical experience, and the design manager may simply disagree with what has been required. If a strong enough argument can be made and submitted to the committee responsible then the standard may be changed accordingly. With this in mind, many companies encourage their design staff to become involved with the committees that develop the standards in their line of business.

Codes (and Codes of Practice) provide general design rules for specific types of product, especially when safety is a key issue, such as with pressure vessels. They often incorporate other standards, and their adoption by regulatory authorities is often used to make compliance mandatory. The basic idea of a Code is to keep designers on the right track by giving them guidance through encapsulated experience. For example, the pressure vessels for the erosion–corrosion reactors mentioned at the beginning of this chapter were originally designed from basic stress equations, not according to the ASME Boiler and Pressure Vessel Code. This is why the top cap in the conceptual design (Figure 8.2) looks so thin. Maybe the vessel would hold the pressure safely with such a thin cap, and in fact both pressure vessels passed their first round of tests with blank caps to that design. However, once holes are drilled for fittings, the bolts are not evenly torqued, minor operational damage or degradation occurs, seal surfaces become marked or pitted, and a whole host of other engineering factors are taken into account then it becomes apparent that a thicker cap is essential. How thick is thick enough? Calculations according to the ASME Boiler and Pressure Vessel Code Section VIII, Division 1 rules are based on simplified empirical equations that result in a conservative design considered safe for all operating conditions. Division 2 rules require a more detailed analysis of the specific design to establish a better understanding of the vessel's engineering performance, and then allow higher stress levels to be used, which results in a less conservative design.

Some guidelines for the use of Standards and Codes:

- Use them whenever appropriate. They can save money, time, and arguments.
- Check which standards and codes apply to the product or system being designed. There may be several and they may conflict.
- Obtain the latest edition of any standard or code to be followed.

- Meet the requirements as closely as possible (even if they are not mandatory).
- Safety standards take priority over rationalization procedures and economics.

8.4.4 Purchased Components

Generally, it makes economic and practical sense to buy ready-made assemblies and components whenever possible, and the main thing to ensure is that the item bought meets the specification stated. For example, many of the quality problems with Jaguar cars manufactured during the 1970s stemmed from faulty or low-quality purchased components, and the quality of the cars dramatically improved when Jaguar started billing suppliers with the full cost of claims caused by failure of such components. In the USA, where the home market is huge compared with many other countries, there is more scope for using purchased components. Mail-order catalogs are a way of life. For a very short while, one of us once worked for a Chicago company making a variety of large automatic assembly machines where all the parts were purchased except for specialized operating stations. Machines were designed and built in a maximum of 3 months by teams of two, who were then removed from the machine. A "debugger" was then given 3 weeks to make the machine "work." It was then delivered to the customer. No drawings were ever done (in fact they were banned); there were no operating instructions, no spare parts and no documentation. It was a "mind into metal" approach, which happened to be successful as long as it was confined to the design of machines involving components such as dial plates, rollover cams, and standardized mechanisms.

The following are a few guidelines for the use of purchased components:

- Use them whenever possible.
- Obtain price and delivery quotations.
- Specify functional and quality requirements in agreement with the supplier.
- Inspect components on arrival.
- Insist that components meet agreed specification.

8.4.5 Layouts and Models

Layouts and models are important tools for communication, negotiation, understanding, and development throughout the design process. Until a layout was done, the team designing the gasifier test rig was unable to negotiate for proper space for the control room. Until a model was built, the team designing the Scotts Rotary Lawn Spreader described in Chapter 7 was unable to gain support for the project. It is said that layouts and models externalize personal thinking and facilitate communication between the right-hand and left-hand sides of the brain. Certainly, layouts and models help the design engineer to visualize the complete design clearly.

The following are a few guidelines concerning layouts, drawings, and computer printouts:

- Use standard sheet sizes, or computer-generated format, for all work.
- Label and date all layouts and printouts.
- Use a standard titleblock (or devise one) for all drawings, including name, date, and tolerancing.
- Number all layouts and drawings according to a rational system (*e.g.* by subsystem and sheet size). Allow for addition of new drawings and revisions.
- Choose scales that will assist in building up and checking assembly drawings.
- For mechanical assemblies use a different sheet (and number) for every component.
- Use the smallest convenient size sheets for each drawing.
- Comply with the current national standards.
- Retain original copies in file with revisions shown.

8.4.6 Prototypes and Testing

The question of whether or not a prototype is built as part of the design process depends on what is being designed. In the case of some high-volume products it is now possible to have sufficient confidence in the design to allow manufacture of the first units directly on the production line, whereas in the case of others, specially built prototypes are needed for extensive field-testing before final design. In the case of one-off (or low-volume) products, manufacture of the final product often proceeds directly from design, and a systematic testing and commissioning procedure is generally required before acceptance by the customer. Whichever approach is appropriate for the project in question, there are some useful general guidelines concerning prototypes and testing. The following set has been adapted from project planning rules suggested by The Technology Partnership (1988) in the UK:

- Prove new technologies separately.
- If major tooling is required, prove the design first.
- Test the hardware that is developed.
- Anticipate having to do modifications.
- Expect and allow for integration problems.

Example: Gasifier Test Rig

The developed concept, or embodiment design, of the gasifier test rig is shown in Figure 8.7. By comparison with the reactor concept in Figure 7.8, it will be seen that the design has passed through another development phase. Many different approaches are used for developing concepts, with the one chosen depending on the nature of the project. For the gasifier test rig the approach used was progressive detailing of layouts, rather than prototyping, modeling, experimenting, or computing. In practice, it was found

Continued

difficult to classify work hours specifically as "conceptual design," "embodiment design," or "detail design," but it had to be done in a definite way for analysis; so, all those hours between the meeting when the design specification was finalized and the meeting when the concept was finalized were classified as "conceptual design." Subsequent hours were divided into "embodiment design" or "detail design," depending on whether they contributed to the development of the reactor concept and overall layout (embodiment design) or they dealt with individual components, detail part drawings, or detail calculations (detail design). This proved adequate except for those hours spent on cost-justification documentation and those spent on design of the control system. The cost-justification documents referred to the developed concept, so these hours were categorized as embodiment design. For the control system design, each interchange was considered individually. There were task clarification hours for the contract controls engineer, as well as embodiment and detail design hours, but the conceptual design had been completed previously.

Whereas conceptual design was mainly concerned with the reactor assembly, embodiment design was concerned with the development and integration of all seven sub-systems for the gasifier test rig. Figure 8.7 shows the developed concept of the reactor only. Examples of its features are:

- Sub-assembly cartridge for the specimens and instrumentation modified to incorporate partial separation of tars and gases.
- Heating element cartridge modified to accept four independently controlled elements instead of two.
- Double O-ring seals with leak detection and provision for emergency nitrogen pressurization between them.
- Annular-groove weld preparation in pressure vessel cap to permit the welding of replacement "inner reactor chamber" tubes to this cap with no need for certified inspection.

These features, and the many others like them, may be considered in terms of the embodiment design guidelines recommended by Pahl and Beitz (1984), and may be assessed according to the embodiment design checklist. For example, almost all the guidelines were used in the development of the "inner reactor chamber" welded fabrication, as shown in Figure 8.8. The overall output from the embodiment design phase of the project included: the developed reactor concept; the equipment selection and incidental design for the seven sub-systems; the preliminary and detailed overall layouts; a more detailed cost estimate with cost-justification documentation; and the control system design complete with the P&I diagram. Final layouts produced were well received by the "customer" and "users," and through them the project gained more support at this stage. The quality of output from this phase was considered satisfactory, but productivity was low. Embodiment design took 770 h (35% of the engineering design effort).

Developed Concept: Embodiment Design 167

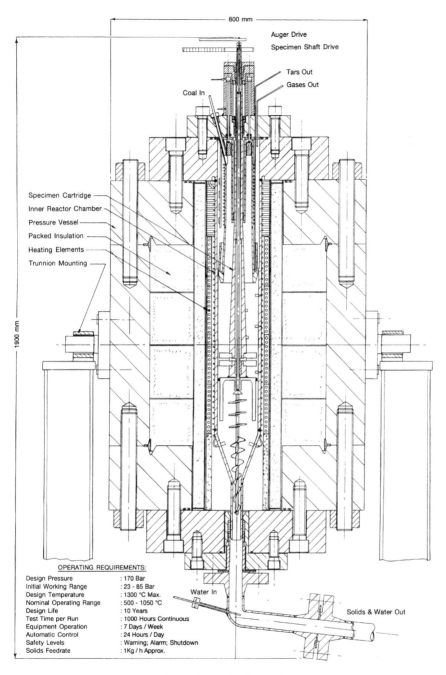

Figure 8.7. Embodiment design of GTR reactor

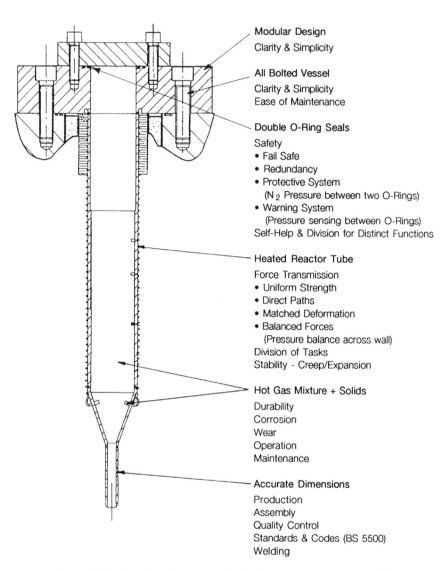

Figure 8.8. Example to show where some of the embodiment design guidelines were applied

8.5 Embodiment Design Checklist and Work Sheet

The *Embodiment Design Checklist* shown in Figure 8.9 lists a set of points for the design manager to consider during reviews of the design work at this stage of development. They are based on the guidelines for embodiment design as described in this chapter and are best used in the form of questions to ask the design team. The *Embodiment Design Work Sheet* shown in Figure 8.10 may then be used to assess the project status and define action items for completion before the design is finally approved for the detailing of all components. Of course, this does not mean that detailing cannot start before this. In fact, the detailing of many standard components may even be complete by this stage. However, it provides the design manager with a final check on important issues concerning the design before it is too late for corrections or modifications to be made. As it is likely that many more hours of design effort will go into this phase than the conceptual design phase, the design manager may find it helpful to use the embodiment design checklist and work sheet more or less continuously to ensure that all aspects of the work are completed in a timely and well-organized fashion. Figure 8.11 provides a worked example for the developed concept for the Life chair shown in Figure 8.12.

The completed work sheet showed a good level of confidence in the *functional requirements* for the Life chair. Contributing factors were:

1. The *overall geometry* of the chair was considered to be good; however, the method of attachment of the back suspension fabric was unresolved.
2. The *motion of parts* factor was considered to be good, *e.g.* the mechanism connecting the seat pan to the back frame ensured that good eye height was maintained as the user reclined in the seat.
3. The team had identified the *forces* involved. An extensive FEA was performed on all parts to check the stresses and deflections. Specialist FEA engineers had also verified this work.
4. The *energy needed* was considered to be good. The chair was designed to be self-adjusting, and its normal operation required a minimum amount of user input and energy.
5. *Suitable materials* had been obtained for all parts. The critical components had been manufactured with final materials and final processes, *e.g.* the seat pan was injection molded using a soft tool. Although it was perceived that the structural properties of rapid prototyped parts would be better in the production version, there was uncertainty over the final cost of these parts.
6. The *control system* consisted of very simple mechanisms, such as the mechanism for linking the seat to the back support. Although the reliability of these control systems was unproven, the team was confident that they would be reliable owing to their simplistic construction.
7. *Information flow* was considered to be good. Instructions for use of the chair were included as a permanent fixture.

EMBODIMENT DESIGN CHECKLIST

REQUIREMENTS	CONTRIBUTING FACTORS	POINTS TO CONSIDER
FUNCTIONAL	Overall geometry Motion of parts Forces involved Energy needed Materials to be used Control system Information flow	Fit of part, assembly of parts, simplicity, clarity Will it work, other functions needed, working principle, division of tasks Strength, stability, stiffness, fatigue life, side effects, flow lines Supply, storage, efficiency, self reinforcing, self help Degradation, wear, corrosion, expansion and contraction Reliability, assembly, testing, trouble shooting Necessary, sufficient, calculations correct
SAFETY	Operational Human Environmental	Safety hierarchy, safe life, fail safe, redundancy, protection, warnings Regulations, standards, codes, history Harmful effects, long term effects
QUALITY	Quality assurance Quality control Reliability	Overall system, life-cycle, standards and codes Manufacturing quality, measurement, monitoring Operation, maintenance, user environment
MANUFACTURING	Production of components Purchase of components Assembly Transport	Can parts be made, layout and drawings adequate Reliable sources, timing, quality assurance, appropriate use Simple assembly, clear sequence Safe internal transport, safe external transport
TIMING	Design schedule Development schedule Production schedule Delivery schedule	Current status, planning, problems Test equipment, test plan, documents, certification Timing, materials supply On time or not, ways of improving
ECONOMIC	Marketing analysis Design costs Development costs Manufacturing costs Distribution costs	Review and update Percent completion, phase diagram, cost overruns Phase diagram, supplier estimates, item by item estimate Tooling cost update, materials additional cost How to be shipped, distribution network
ERGONOMIC	User needs Ergonomic design Cybernetic design	Reliable and easy to use User friendly, good physical layout Good controls
ECOLOGICAL	Material selection Working fluid selection	Source, supply, disposal, mixture, safety Safety, toxicity, replenishment
AESTHETIC	Customer appeal Fashion Future expectations	Survey, comments Competition Reliability of predictions
LIFE-CYCLE	Distribution Operation Maintenance Disposal	Quietness, vibration, handling Simple inspection Simple maintenance, user safety, who does it? Recycle, scrap

Figure 8.9. Embodiment design checklist

Developed Concept: Embodiment Design 171

EMBODIMENT DESIGN WORK SHEET   PROJECT: _____   DATE: _____

REQUIREMENTS	CONTRIBUTING FACTORS	CURRENT STATUS (Good / Marginal / Poor)	REQUIRED ACTION (Proceed / Revise / N/A)
FUNCTIONAL	Overall geometry Motion of parts Forces involved Energy needed Materials to be used Control system Information flow	☐ ☐ ☐ ☐ ☐	☐ ☐ ☐
SAFETY	Operational Human Environmental	☐ ☐ ☐ ☐ ☐	☐ ☐ ☐
QUALITY	Quality assurance Quality control Reliability	☐ ☐ ☐ ☐ ☐	☐ ☐ ☐
MANUFACTURING	Production of components Purchase of components Assembly Transport	☐ ☐ ☐ ☐ ☐	☐ ☐ ☐
TIMING	Design schedule Development schedule Production schedule Delivery schedule	☐ ☐ ☐ ☐ ☐	☐ ☐ ☐
ECONOMIC	Marketing analysis Design costs Development costs Manufacturing costs Distribution costs	☐ ☐ ☐ ☐ ☐	☐ ☐ ☐
ERGONOMIC	User needs Ergonomic design Cybernetic design	☐ ☐ ☐ ☐ ☐	☐ ☐ ☐
ECOLOGICAL	Material selection Working fluid selection	☐ ☐ ☐ ☐ ☐	☐ ☐ ☐
AESTHETIC	Customer appeal Fashion Future expectations	☐ ☐ ☐ ☐ ☐	☐ ☐ ☐
LIFE-CYCLE	Distribution Operation Maintenance Disposal	☐ ☐ ☐ ☐ ☐	☐ ☐ ☐

Figure 8.10. Embodiment design work sheet

Embodiment Design Work Sheet

PROJECT: LIFE CHAIR **DATE:** JUNE 2001

Requirements	Contributing Factors	Current Status (Good / Marginal / Poor)	Required Action (Proceed / Revise / N/A)
FUNCTIONAL	Overall geometry	Marginal	Proceed
	Motion of parts	Good	Proceed
	Forces involved	Good	Proceed
	Energy needed	Good	Proceed
	Materials to be used	Marginal	Proceed
	Control system	Good	Proceed
	Information flow	Good	Proceed
SAFETY	Operational	Good	Proceed
	Human	Good	Proceed
	Environmental	Good	Proceed
QUALITY	Quality assurance	Good	Proceed
	Quality control	Poor	Revise
	Reliability	Poor (marginal/poor)	Proceed
MANUFACTURING	Production of components	Marginal	Proceed
	Purchase of components	Marginal	Proceed
	Assembly	Good	Proceed
	Transport	Good	Proceed
TIMING	Design schedule	Marginal	Revise
	Development schedule	Marginal	Revise
	Production schedule	Poor	Revise
	Delivery schedule	Marginal	Revise
ECONOMIC	Marketing analysis	Good	Proceed
	Design costs	Marginal	Revise
	Development costs	Marginal	Proceed
	Manufacturing costs	Marginal	Proceed
	Distribution costs	Marginal	Proceed
ERGONOMIC	User needs	Good	Proceed
	Ergonomic design	Good	Proceed
	Cybernetic design	Good	Proceed
ECOLOGICAL	Material selection	Good	Proceed
	Working fluid selection	Good	N/A
AESTHETIC	Customer appeal	Good	Proceed
	Fashion	Good	Proceed
	Future expectations	Good	Proceed
LIFE-CYCLE	Distribution	Good	Proceed
	Operation	Good	Proceed
	Maintenance	Good	Proceed
	Disposal	Good	Proceed

Figure 8.11. Example of embodiment design work sheet

Figure 8.12. CAD model showing the embodiment design solution for the Life chair. Courtesy of Formway Design

Safety and compliance with safety standards were considered to be critical to the success of the Life chair project. Contributing factors were:

1. The *worst-case load* scenarios had been thoroughly analyzed and tested for. The team was confident of the operational safe-life. In cases where failure had potential to result in injury (*e.g.* failure of the back frame), extra redundancy was included in the structure.
2. The team had been thorough in ensuring that they met the *applicable safety standards*, and the human influence was positive.
3. *Environmental safety*, in this case, concerned the chair's ability to provide long-term support to the human body safely. The team considered that the chair delivered a better level of long-term support than any of the competitor products.

Factors contributing to *quality* included:

1. *Quality assurance* was tested against relevant industry standards. The chair was found to perform to the highest level in all cases.
2. *Quality control* factors had not been considered. The design team realized that this was a weakness and they were preparing to put a quality control strategy in place.

3. *Reliability* had been established to an acceptable level for the prototype. However, the team was not confident about the product's performance in the user environment.

Contributing factors for the *manufacturing* requirements were:

1. The team was happy with the *production of components* influence: components could be manufactured and controlled at a production level. There was uncertainty, however, regarding larger production runs; the team members were not experienced at manufacturing mass-production quantities.
2. Reliable *suppliers* had been identified for the prototyping phase and many had supplied components for batch production runs of other furniture. There was uncertainty regarding the purchase of components factor because of the unproven ability of suppliers to meet the demands of larger production runs reliably.
3. The *final layout* was elegant in terms of design for assembly. This was attributed to the use of modular components requiring a minimum number of assembly operations. Many of the components could be assembled without tools, and this had a positive influence on manufacturing requirements.
4. The *transport* influence was positive because the product could be safely handled both within the production process and externally. The chair design allowed for separation into efficient packaging elements that could be easily reassembled when the product reached its final destination.

Factors contributing to the *timing* requirements were:

1. The final layout details took longer than planned and the *design schedule* had not been met. This delayed negotiations with potential manufacturing partners and had a negative influence on the development of the chair.
2. The *development schedule* allowed time for final testing; however, the current status was marginal due to insufficient contingency for unplanned events.
3. *Production schedule* had not been considered; this factor was to be revised once a partner had been found.
4. The *delivery schedule* for the final working prototype was unrealistic and needed revision.

Drivers for *economic* requirements were that the product be cost competitive and be considered value for money. Contributing factors were:

1. The *marketing analysis* for the local market was thorough and had involved feedback on the prototype from preferred customers. The US market needed to be reviewed.
2. *Design costs* were still within budget; however, additional design contracting staff were required to complete the embodiment phase. The current status was marginal–good.

3. *Development costs*, though expensive, were within budget because only one set of soft injection-molding tools had been acquired when multiple sets had been budgeted for.
4. Projected *manufacturing costs* were conservative; however, the teams inexperience at mass production created uncertainty.
5. *Distribution costs* were uncertain because the Formway team had no established distribution network in the USA. Despite this, they took as many steps as they could to make cost predictions.

Ergonomic factors were a focal point for the Formway design effort. In terms of meeting the user needs and from an *ergonomic design* perspective, the chair performed better than hoped and out-performed competitor products. Considering *cybernetic design* factors, a minimum number of user control inputs were needed and the control devices were easy and obvious to use.

Ecological factors were strongly positive. Materials selection was from approved suppliers and a minimum quantity of materials had been used.

Aesthetic qualities were a primary driver for the Life chair project. Initial reaction from customers indicated a strongly positive customer-appeal factor. Customers liked the minimalist aesthetic look, which was also positive from a fashion perspective because it followed current trends. This meant that the product was likely to be stable in the marketplace. The *future expectations* factor was good. This was because the demands of different users could be met by either dressing the chair up or down with different fabric qualities or by adding or removing component blocks, such as the arm rests.

Factors influencing the *life-cycle* requirements were:

1. *Distribution* had been planned for in the local market; however, distribution remained unresolved for Europe and the USA.
2. *Operational factors* were good because the layout allowed simple inspection.
3. The *maintenance factor* was good; the chair was designed for easy assembly and disassembly, allowing straightforward replacement of components.
4. *Disposal factors* had a good influence on life-cycle requirements. The prototype chair was made from like families of recyclable materials. Contamination had been reduced by using natural finishes (*e.g.* polished aluminum) and by avoiding the use of composite materials.

The completed work sheet shows a good level of confidence in terms of the functional, safety, ergonomic, and aesthetic performance of the prototype. Areas of weakness are quality, timing and economic factors. These weaknesses were generally associated with the fact that the Formway team was developing a product to be sold at much larger production volumes than they were familiar with.

8.6 Tips for Management

8.6.1 Overall Product

- Test functional performance — must meet or exceed customer expectations.
- Check for economic feasibility — cost must be acceptable to the customer.
- Check safety performance — must meet applicable safety standards.
- Test ergonomic performance — user expectations must be satisfied.

8.6.2 Overall Design

- Insist on clarity, simplicity, and safety in the design.
- Ensure that force transmission paths will be satisfactory.
- Ensure satisfactory allocation of functional tasks to components.
- Ensure that self-help has been appropriately incorporated.
- Ensure that self-damaging effects have been investigated.
- Ensure that the design will perform in a stable manner.
- If planned instability is used, ensure required effect will be achieved.

8.6.3 General

- Ensure that calculations are appropriate, adequate and correct.
- Ensure that calculations have been recorded in a professional manner.
- Ensure that the most appropriate materials have been selected or designed.
- Ensure that the requirements of applicable standards and codes are met.
- Ensure that purchased components have been incorporated effectively.
- Ensure that models, layouts, printouts, and drawings have been used to best advantage.

Chapter 9
Final Design: Detail Design for Manufacture

9.1 The Importance of Detail Design
9.2 The Design Manager and Detail Design
9.3 Quality Assurance
9.4 Interaction of Shape, Materials, and Manufacture
9.5 Manufacturing Drawings and Information
9.6 Standard Components
9.7 Assembly
9.8 Testing and Commissioning
9.9 Detail Design Checklist and Work Sheet
9.10 Tips for Management

9.1 The Importance of Detail Design

One of the most difficult things for a design manager to get across to corporate management, the design team, and customers alike is the importance of small details in design. There is a tendency to leave the details to a student assistant, the draftsperson, or perhaps the computer! This is a serious mistake. Most accidents and disasters involving engineering design issues can be traced back to errors, inexperience, or poor judgment in detail design. We have personally investigated many fatal or serious injury accidents in which simple deficiencies in detail design were the direct cause or a major contributor. A few examples are:

- Omission of a plastic locking plug on a 3/16-inch diameter screw thread (fatal truck collision).
- Internal corrosion of a steel box-section suspension arm (vehicle rollover).
- Threading a load-bearing 4 mm machine screw into a sheet metal hole (loss of one eye).
- Failure to remove feathering and burrs from a stainless-steel tray (severed nerves in a finger).
- Failure to provide anchoring for the fixed end of a cantilevered step (slip and fall back injury).
- Use of a cattle clasp as a tow hook on an equipment tether (fatal pipeline accident).
- Spring-loaded cable anchored to easily removable door bracket (loss of one eye).

Each one of these accidents could have been prevented by more care in carrying out or checking the detail design, without needing any engineering analysis, calculations, or sophisticated knowledge. It is the responsibility of the design manager to ensure that sufficient emphasis is placed on excellence in detail design, to make sure that there is competent staff to carry it out, to check carefully for possible problems, and to warn of the dangers if detail design is not treated with sufficient respect.

9.2 The Design Manager and Detail Design

A moment of truth for the design manager comes when a design is approved for manufacture. Now what was on paper or in a computer is going to become a reality, at great cost in materials and manufacturing effort. There is no turning back. Every single component in every single sub-assembly in every single unit of the system is going to have to be made or procured, then fitted together with all those other components that have never been together as a whole before. There are bound to be a few sleepless nights before the design "works" as intended. The problem is that, with any kind of sophisticated piece of equipment, the design manager cannot personally check every drawing or calculation that has been done. There may be hundreds, if not thousands, of documents or their equivalent, and the interrelationships between them can be extremely complex. The design manager, therefore, has to develop a "nose" for potential problems and be on the lookout for them throughout the detail design phase, not just in a final review. The idea behind this chapter is to help the design manager develop a systematic thinking process, which will help ferret out detail design problems before they cause serious mischief.

In passing, there is one simple technique that one of us has often used on design projects coming close to manufacture. Perhaps it arose out of needing something more challenging than counting sheep during occasional sleepless nights. In your mind's eye, think of all the components that have been detailed, visualize their manufacture, then assemble the whole system, component by component. Continue by thinking of a user starting up the equipment or product and think through different ways of operating and maintaining it right to the point when it is worn out. Maybe you nod off before getting very far along in this thinking process, but the exercise is systematic, it keeps the thinking going at night, and it can also help you to get to sleep!

9.3 Quality Assurance

"Quality assurance is an all-embracing term that covers every activity and function concerned with the attainment of quality" (Morrison, 1985). As stated in the Scott Fetzer Quality Improvement Guidelines: "Quality means meeting customer requirements and exceeding customer expectations... The customer includes the end users of products and services as well as other Scott Fetzer employees" (Birmingham, 1991).

As discussed in Chapter 2, with reference to Figure 2.1, *quality assurance* must be considered as an integral part of a design project and not something that can just be added in at the end. Quality assurance requires specific *quality management* as prescribed, for example, by the ISO 9000 series of International Standards for Quality Management listed in Chapter 11. The implementation of quality management may often be enhanced by the application of specific tools such as Taguchi methods and QFD already discussed in Chapter 3. This is mentioned again here to highlight the need for the design manager to work particularly closely with the quality assurance team during the final phase of design before manufacture. By use of the more sophisticated approaches now becoming accepted, it becomes possible to combine the quality assurance and design functions into a fully integrated *quality engineering* approach for some types of design project (Clausing, 1994). The design manager needs to work towards quality improvement on a continuing basis and should be involved directly with the implementation of quality improvement measures. A useful source of information on this issue is the American Supplier Institute in Dearborn, Michigan, which has published many helpful techniques, guidelines, and case studies in various forms, such as on Taguchi methods (ASI, 1990; ASI/JSA, 1990).

Example: Electronic Deadbolt for Locking Doors
Courtesy of Manz Engineering Ltd.

For many years, a unique electronically controlled dropbolt (deadbolt) system for the remote locking of building doors has been manufactured by a small family business in New Zealand. The complete unit fits inside the doorframe mortice, with the cylindrical deadbolt solenoid operated through a mechanical linkage and powered by a 12–24 V electrical supply. Alignment of the bolt and striker plate is sensed magnetically during operation, and an electronic circuit board provides monitoring, operational, and safety capabilities. From its origins as a simple electrical device, incremental detail design developments have transformed it into an exceptionally high quality, robust, and reliable product, marketed internationally through major suppliers to the security industry. Each unit is assembled, tested and packaged by hand, and most of the parts are machined in-house or manufactured locally.

One day a large order was received from a multinational electro-mechanical security equipment corporation based in Germany. This sudden demand caused a scramble in the four-person company, so as to deliver the first batch of 500 units on time. Then, unexpectedly, their

Continued

product was rejected for not meeting the customer's quality assurance requirements. The entire shipment was returned on a single massive pallet. This was a devastating blow to a tiny, specialty manufacturing company, but it accepted the loss and investigated the problem. They heard that the corporation sales and marketing staff had seen a sample lock at a trade show and were impressed enough to order the units through the purchasing department without the necessary engineering approval. The engineering department rejected the units for several valid reasons, including:

- Chatter marks on the countersunk screw holes in the faceplate.
- Use of "flying leads" for electrical supply wiring, instead of connectors.
- Life tested to only 200 000 cycles instead of the required 1 000 000 cycles.

The machining tolerances on the faceplate countersunk holes were revised and samples sent to define the tolerance band on diameter and depth. Connectors were fitted to the circuit board and a plastic injection-molded component was modified to maintain the same overall dimensions (involving the high cost of adding material *back* into a die cavity). Life tests revealed some wear in the mechanical linkage above 300 000 cycles with normal lubricants, but this was eliminated by using ROCOL J166 anti-seize grease. Finally, the product met all the customer's requirements and was accepted as their first product line to be manufactured by an outside company.

Although this was a costly exercise, the manufacturing company regarded it as a beneficial experience, as it resulted in a much-improved product, regular orders, and a long-term working relationship, with the following lessons learned:

- Always obtain written design specifications from the customer.
- Notify the customer of any component or other changes in advance.
- Send samples and obtain approval for any changes, prior to manufacture.

9.4 Interaction of Shape, Materials, and Manufacture

Each individual component has a certain geometrical *shape*, is made from a particular *material*, and is made by a particular *manufacturing process*. These three things are important in themselves, but the interrelationships between them are equally important.

The shape of a component is generally set by spatial and functional requirements, modified by constraints such as weight or interference with other components. The dimensions and description of surfaces on a drawing commonly define the shape, and this is an area where the computer has long been of assistance in design.

It used to be that a material was selected for a particular component from a limited range available. Great advances in materials science have changed this in a number of ways. The range of materials available has increased, with performance characteristics more precisely controlled and quantified. In many cases it is now possible to design a material for the specific application, which means addressing the issue of materials right from the beginning of the design process. Certainly, the designer has a much better range of materials to choose from; but the selection process is more complex, and the way the material is processed and handled during manufacture has to be considered in detail. For example, consider the feedwater lines for high-pressure power boilers. Although the ASME code specification for the pipe steel composition has remained essentially the same for decades, the carbon steels manufactured today that meet the same specification in fact have slightly different characteristics than those manufactured years ago. The composition of the SA-106B steel is specified as: carbon 0.3%; manganese 0.29%; phosphorus 0.048% max.; sulfur 0.058% max.; and silicon 0.10% min. There is no exclusion on having, for example, a small percentage of "tramp metal" such as chromium, in the steel; indeed, when pipe material that was installed in the mid-1900s is analyzed it is common to find a small percentage of chromium present. The current steels are much "cleaner" because of improved steel-making technology, and the expensive chromium element is unlikely to be present in this particular steel, where it is not a required part of the composition. The presence of the "tramp" chromium actually provided a certain level of protection from corrosion, a fact that was not recognized until quite recently when some serious feedwater line explosions led to detailed failure investigations. It is now becoming a practice to specify a low-alloy steel such as SA-335 for these feedwater lines, with 1.0 to 1.5% chromium and 0.44 to 0.65% manganese to help protect against flow-accelerated corrosion and flow-assisted chelant corrosion (Hales et al., 2002).

Product manufacturers also have to adapt when faced with designs specifying unfamiliar materials. A company that had an excellent reputation for high-quality work on tooling for the manufacture of Rolls Royce aircraft engines changed from traditional tool- and die-making to the manufacture of end-effectors for robots due to economic pressures. All of a sudden the customers began to reject their work. It transpired that while the tool-and-die business primarily involved the use of hardened steels, the end-effector business primarily involved the use of softer aluminum alloys. The toolmakers were damaging their own products by, for example, sliding them across a metal-clad bench or nicking them during assembly. The whole company had to be re-educated in how to design, manufacture, and protect assemblies involving aluminum parts.

Often, it is not only the physical and mechanical properties of a material that are important. The thermal, electrical, and chemical properties may need to be taken into account in case of peculiar or adverse effects. Heat treatment and special surface treatments are commonly used to enhance the properties of a material after machining, and the specifications for such treatments are becoming more stringent. If the process is not controlled carefully enough then the material may not meet the physical characteristics the designer has assumed for calculating strength, wear, and fatigue performance. Service failures could occur. Integrated suites of computer programs, such as the Cambridge Engineering Selector, CES4 (Cebon, 2003), are now available to assist in the specification, selection and design of materials, taking into account the production processes and heat treatments. The Cambridge Engineering Selector was developed specifically from the designer's point of view, rather than from the materials science point of view usually encountered with earlier materials database programs.

With regard to manufacture, it is essential for the design manager to team up with the manufacturing staff and to work with them directly. The intent of the design engineers must be fully communicated and understood by those manufacturing and assembling the parts. For the design of mass-produced products, the techniques advocated by Taguchi (Clausing, 1994) have been developed to help make this link effectively and efficiently. Manufacture today is a closely controlled process in almost every field, and the manufacturing process to be used greatly affects the design of a product. The quality of the components manufactured is also critical. If the components are not made as designed then there are bound to be problems. As previously mentioned in another context, one of us investigated design problems in a case where an extremely large prototype tri-axis transfer press failed to perform to specification. The contract had been set up in such a way that there was no interaction between the design team in one company and the manufacturing team in another. To complicate matters further, one company was in North America and the other was in Europe. Language difficulties created misunderstandings through memoranda, and when the drawings were translated the intent of the designer was not fully communicated. Mixed imperial and metric units were used, which had to be rationalized. Material specifications were not exact equivalents from one country to the other, and it was not clear that specified surface treatments and heat treatments were ever carried out according to the original drawings. Approximately 60 000 h of engineering design work was involved and some 4000 manufacturing drawings were shipped from the designing company to the manufacturing company. The customer ended up with a machine that could work only at low speed, so they would not pay the manufacturer, and the manufacturing company would not pay the designing company on the basis that the design was faulty. The cost of claims and counterclaims far outweighed the project cost, and most of the problems could have been avoided had there been closer communications between the two companies during detail design.

9.4.1 Shape–Material Interaction

An overall guideline of particular importance to the design engineer is to design a component so that it can be made as closely as possible to its final shape ("near net shape") by means of a primary process. In other words, to be made in "one bang" if possible. Every secondary forming operation and every machining operation costs money and increases waste. A hollow shaft is made more economically from a pre-finished tube than from a solid bar, and perhaps the design can be adjusted to accommodate this. Components must be strong enough to carry the loads imposed during operation. But how strong is strong enough? This depends on the nature of the component and the risk associated with any potential failure. It may depend on deflection, fatigue life, corrosion resistance, strain hardening, stress-corrosion resistance, or numerous other factors. Part of detail design is to assess what type of failure might occur with a particular component and to design to an acceptable level of resistance to such failure. A safety factor may be introduced to account for uncertainties in the analysis, with the size of safety factor depending on the cost, feasibility, and benefits of doing more detailed analysis. For example, the ASME Boiler and Pressure Vessel Code, Section VIII, has two sets of rules: Division 1, based on simplified assumptions and calculations using a general "safety factor" of four, and Division 2, which uses a less conservative "safety factor" but requires far more detailed calculations. If the vessel you are designing is relatively small and will operate at low pressure, then the cost of performing the additional analysis probably would not be offset by any reduction in material thickness of the pressure-vessel components. You may not even be able to take advantage of an allowed thinner wall if it does not calculate to a standard thickness. However, if you are designing a large high-pressure vessel, as for a coal-gasification plant, then the reduction in wall thickness permitted through use of Division 2 rules can result in major cost savings. Reductions in thickness may even enable the use of a less expensive manufacturing process for the components.

Another point to consider is that a material may be strong enough in the sense that the calculated internal stresses are low, yet it may fail in its intended purpose because of excessive deflection when the external forces are applied. An example of this involves the handrails around scaffolding. Scaffolding safety standards often require that handrails accept a certain horizontal force without exceeding the yield strength of the material or causing permanent deformation. However, if this is the only criterion used, then it is quite possible to design handrails that meet the standard yet deflect so much that they would fail to restrain a person from falling overboard.

A few general hints with regard to shape-material interaction are as follows:

- Use axial forces whenever possible.
- Concentrate material away from the neutral axis in bending.
- Beware of high stresses in bending.
- Complementary shear stresses can be used to advantage in design of thin sections.
- Thin-walled closed sections are good for torsion.

9.4.2 Shape–Manufacture Interaction

The shape of a component may look fine on a drawing, but it could create all kinds of problems for manufacture. For example, a new design of timing gear for a truck engine initially had four windows in the web instead of three as previously designed. The manufacturing engineers asked the design engineer why it had four windows instead of three. There was no particular reason, except that it seemed to look better. Whereas the previous gears could be finish-turned in one chucking operation using a self-centering three-jaw chuck, with four windows in the web they would have to use an independent four-jaw chuck, requiring much more set-up time. They settled for three windows.

Continuing with the same example, it turned out that the supplier of the forging blanks had recently moved operations from Canada to Italy, and since that time 90% of the gears needed balancing after machining instead of only 10%. Even though the forgings were within tolerance dimensionally, the shape of the webs varied sufficiently to cause balancing problems. The design engineer was asked to specify closer dimensional and geometrical tolerances on the new design to fix the problem. Specifying tighter tolerances is generally considered to increase costs; but, as this example shows, it depends on the circumstances. The way to make a proper assessment is to talk with the manufacturing staff. There is no sense in struggling to widen tolerances on a part that for other reasons would be machined on a jig-borer anyway. The part will not be any cheaper. However, if by widening a tolerance the part could then be simply milled, then it would be worth considering. It must be very frustrating for a manufacturing engineer to put a lot of effort into meeting precisely what was asked for on a drawing, only to find out later that constraints that made things tricky were fictitious to start with. Geometric tolerances need careful thought, as they can create unexpected problems for the manufacturer. In one case the designer had defined a reference datum that could not be measured from on the actual part. The manufacturer set up an equivalent datum that could be measured from, but this introduced a sufficient difference in the geometric tolerance stack-up that a large casting had to be scrapped after machining.

Surface finish, residual stresses, and flaws in the material are important when it comes to detailing, and the design manager is often called on to make quick decisions when these cause problems during manufacture. An additional factor in the gear example above was the distortion that occurred during machining of the forged gear blank due to relief of residual stresses, and the design engineer was also asked to consider this aspect in the new design. Sometimes a decision must be made on whether to accept or reject a part that has had expensive machining carried out prior to finding a flaw in the remaining material. The manufacturing engineer may want to press ahead, making minor dimensional changes so that the flaw can be machined out. Certainly, the supplier will make light of it and upper management will not want any more delays. There is a strong temptation to go ahead, and if long lead times or crippling costs are involved then it may be the only way. However, the design manager takes a big risk in agreeing to accept flawed material or changing dimensions to accommodate defects. As a general rule, approval should be given only after a thorough review of all the design implications of the change. It is easy to overlook or forget one or two of the reasons why a particular dimension or tolerance was set the way it was, and there is no "comeback" once the changes have been approved. A typical case was when one us had ordered a pressure vessel with a forged shell in the form of a 724 mm diameter nozzle, 108 mm wall thickness and 610 mm long, for continuous high-temperature operation at over 200 atm hydrogen pressure. Delivery was long overdue and things were getting desperate when the manufacturer called to say that they had just made a mistake in cutting the end off the forging. Somehow, it ended up being cut on an angle, so that the nozzle was too short on one side. We could have a short nozzle, a nozzle built back up with weld, or a new one at some distant time in the future. The tight project deadline left no choice but to build up the wall with weld and then re-machine.

However, before approval was given to the manufacturer, a design exercise was carried out to see what the implications would be, and to work out an acceptable procedure that we were confident would meet the ASME code requirements. In the end the vessel was assembled according to the drawings and was still in operation 12 years later.

9.4.3 Material–Manufacture Interaction

It is common for materials to have their characteristics changed as part of the manufacturing process, and the designer has to be sufficiently knowledgeable in materials science to understand what can or cannot be done and what the implications are. Heat-treating to harden or soften metals is an obvious example. As the design and use of materials gets more sophisticated, so does the interaction with manufacture, such as with the use of high-strength, low-alloy steels in car bodies. It seems a good idea to use stronger steels and thus reduce weight in car bodies, but this creates secondary problems that must then be resolved in their own right. There are springback problems in dies, and deep drawing is difficult. The thinner material "oilcans" more easily, and loss of thickness from physical damage or corrosion is more critical. The sheet may have to be protected by galvanizing or other coating, which introduces additional manufacturing processes. These were important issues at the time when these steels were introduced into car-body manufacture, and it took a lot longer than expected to reduce the weight of vehicles without compromising safety or operational life.

Example: Gasifier Test Rig

Detail design theory draws together techniques used in the "form" design of individual components and guidelines for completing and checking the final production documents. Form design is concerned with the interactions between shape, materials, and the manufacturing process for components, and the integration of components into assemblies. The output from the detail design phase has traditionally been in the form of detail drawings, but it is now often in the form of digitally stored manufacturing information. For the gasifier test-rig project there were no facilities for "computer-aided drafting" available at the time, and all drawings were produced manually. Although there was overlap between the embodiment design and detail design phases, there was a precisely defined point at which detail design started. This was a meeting with the design office manager to agree on a schedule, starting from that day, for the completion of all necessary manufacturing drawings. It marked a definite change of emphasis on the project. Had everything gone as planned, the drawings would have been completed within the time limit set. However, despite the careful planning, no qualified detail designer was available until well into the agreed period. This delayed the work for 5 months.

The "inner reactor chamber" welded assembly, described previously in Chapter 8, provides a typical example of a shape–materials–manufacture interaction problem tackled during detail design, as shown in Figure 9.1. It included: selection of materials; use of the pressure vessel codes BS 5500 and ASME Section VIII; dimensional and geometrical tolerancing; welding sequences during assembly; selection of standard O-rings using the manufacturer's guidelines; and questions of thermal expansion, creep, and heat transfer.

The output from the detail design phase, up to the cut-off point for project data collection, included:

- 42 pages of pressure-vessel calculations;
- 8 pages of scrubber calculations;
- 19 pages of steelwork calculations;
- 18 pages of other calculations;
- 65 detail drawings;
- 14 files of supplier information with index.

Although the work completed was assessed as satisfactory, the productivity during this phase was poor due to sporadic drafting assistance. Detail design took 875 h (40% of the overall engineering design effort), excluding the estimated hours for completion of the drawings after the cut-off point for data collection.

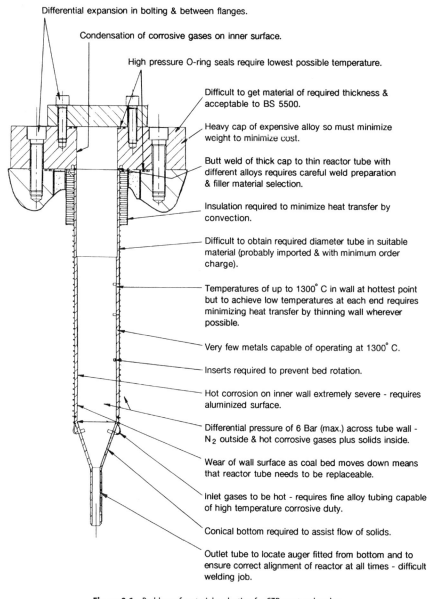

Figure 9.1. Problem of materials selection for GTR reactor chamber

Example: Seat Pan Detail for the Life Chair

The interaction between *shape*, *materials* and *manufacture* required significant consideration and resulted in considerable rework in the detailed design of the Life office chair. Although the team had built a prototype that had exceeded all expectations, it was the final detailing that would be critical in ensuring the introduction of a reliable and cost-competitive product to the market.

The cost of the material used to make the seat pan for the prototype was found to be too expensive for mass production. A more cost-effective material was found; however, this affected the interaction between *materials* and *shape*. The structural properties of the new material were different, and this required changes in size to provide the required flexibility, and hence comfort, while ensuring adequate strength. A more detailed stress analysis was performed at the detailed phase to reduce stress concentrations.

Interactions between *shape* and *manufacture* for the seat pan involved the detailed specification of tolerances on dimensions and tolerances on surface finish for the injection-molding die. These details produced a good fit between the seat pan and its undercarriage, along with the required surface finish.

The interaction between *material* and *manufacture* was governed by the flow of materials in the injection-molding process. The compatibility of the material with this method of manufacture was critical in achieving the best quality seat pan. To refine the final seat pan solution, the specific material properties, along with the die geometry (including positions of runners and gates), were input into a mold flow analysis program. This analysis was used to specify the optimal injection-molding parameters, such as pressure, temperature, and flow rate and runner geometry. Figure 9.2 shows one of the outputs from the mold flow analysis for the seat pan. See also color plate provided.

Color Plate

ENGINEERING POLYMERS APPLICATION DESIGN

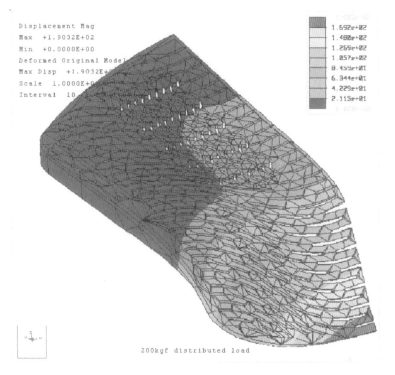

Figure 8.6. FEA plot for the Life chair seat pan. Courtesy of Formway Design

Color Plate

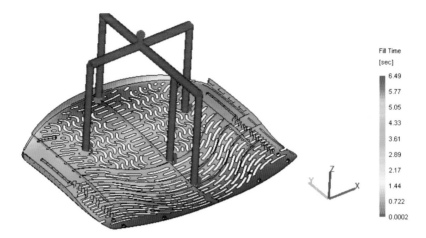

Figure 9.2. The mold flow analysis for the seat pan on the Life chair. Courtesy of Formway Design

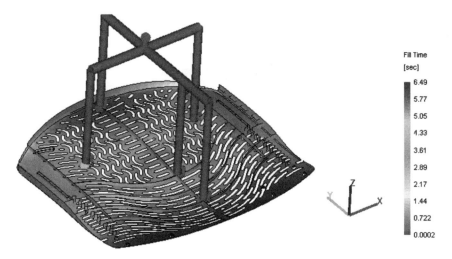

Figure 9.2. The mold flow analysis for the seat pan on the Life chair. Courtesy of Formway Design. See plate section for color version.

9.5 Manufacturing Drawings and Information

Every detail of the design engineer's final design must somehow be communicated to the manufacturing department with sufficient clarity for all the parts to be made and assembled correctly. Whether this is done through the use of computers or drawings on paper depends on the circumstances, but in the end it is the quality of the design engineer's thinking that will determine the quality of the final product most. A poor concept cannot be made good by excellent detail design, and an excellent concept will be destroyed by poor detail design. To be successful, the design manager must be capable of eliciting a winning concept from the design team, then nurturing the idea right through to the very end. At the detail drawing phase this can get very tedious, and the work is totally different from the early days of the project. It is literally a matter of going through every single component drawing to make sure that it is done, it is adequate, there are no mistakes, the material is available, the component can be made, and the component will do its job.

Dimensional and geometric tolerances need to be checked. For example, is the reference surface used as a datum for the geometric tolerancing actually a surface from which measurements can be made in the shop, and is there a way of measuring what is asked for? Are the tolerances fighting each other?

9.6 Standard Components

There is a huge selection of standard components available to the design engineer. The main difficulty in using standard components is sifting through the information to find what is wanted. However, databases for helping with this now exist, and search engines on the Internet, such as "Google," are fast becoming the best way of finding available components. Certainly, there is no excuse for spending design time redesigning what has been done many times before. The things to concentrate on are efficiency in sourcing components and effectiveness in obtaining and maintaining adequate quality of the incoming items. In the USA, which has a long history of excellent mail-order service, it is sometimes possible to do a complete design project using items from a single catalog, as discussed in Chapter 8. This may not sound very elegant to the purists, but it is an effective way of getting something put together in a short time, and the following are some other advantages of using (on-line) catalog items wherever possible:

- Performance and dimensions specified.
- Price accurately known.
- Delivery controlled.
- Can be returned if necessary.
- Illustrations and array of options encourages creative thinking process.

9.7 Assembly

Although the design team may have assembled a prototype and may have a clear idea in principle of how an item of equipment or a product should be assembled, in practice assembly is a complicated business. Again, the manufacturing department should be involved at an early stage in design. Parts must be stored, handled, joined, fastened, adjusted, and inspected during the course of assembly. Texts are available which cover this in detail, and only a few pointers for the design manager will be offered here.

Storage is costly and fraught with unexpected problems. The parts get lost, the parts get stolen, the parts get damaged, the warehouse floods. There is no end to the tale of woe. As design involves something new, it is common for peculiar packages to start arriving at shipping and receiving with no home to go to, no name attached, and no explanations. It is up to the design team to negotiate storage space and handling for equipment, material, and parts to do with the project, then to oversee their disposition on arrival. Perhaps the manufacturing department will take care of it or help. Perhaps there is an empty room somewhere, or perhaps the company is well enough organized to have special procedures for new hardware. In any case, things must be planned out ahead of time, approvals obtained for the handling and storage of parts, and warning given of their imminent arrival.

If large equipment is involved then the problems multiply. Cranes are needed, the thing is bigger than expected, the crate will not come apart, the crate broke open on the journey, and everyone stops work to watch. It is always interesting when something new comes in, but it can be extremely harrowing for the design manager unless the event is well planned in advance. Author Hales was design manager for a test facility involving a series of large pressure vessels, compressors, generators, and other equipment. Almost every major delivery caused uproar in the company because of the disruption and difficulties with maintenance and buildings staff.

9.8 Testing and Commissioning

It is not often realized how much time and cost is involved in the testing of new products or in the commissioning new equipment. The design manager must realize that, for any projects with characteristics high on the Rodwell scales of magnitude, novelty, or complexity (see Chapter 3), careful planning and budgeting will be required for testing and commissioning. In the case of a commercial or consumer product, extensive field tests may have to be carried out to confirm the functional performance and the operation of safety systems. Mandatory testing may have to be performed by designated organizations such the Underwriters Laboratory in the USA before the product can be approved for sale on the open market. In the case of large one-off machines or equipment, it is common for functional testing to be required on assembly at the manufacturer's plant, followed by extensive commissioning trials after final installation at the customer's facility. In extreme cases, the cost of commissioning may be as high as 25% of the total cost of the project.

If testing and commissioning are not accounted for realistically in planning the project then there will be insufficient time and money to prove the product or shakedown the equipment when the time comes. There is a high risk that the project will fail, even if the design is satisfactory. This was a major cause of machine failure in the case of the tri-axis transfer press, used as an example in Chapter 3. It is important that testing and commissioning is carried out in a systematic fashion, and the design manager needs to become familiar with the methods and techniques available appropriate to the particular project.

Example: Low-cost Lawn Spreader

The Scotts™ EasyGreen Model ER-2A Rotary Spreader is one of several well-known lawn spreaders available to consumers in the USA. It is sold either packed in a box disassembled or assembled by the retailer at a slightly higher cost. It so happened that one of the authors of this book needed a small lawn spreader with good distribution control for use on narrow lawn areas between flowerbeds. The various demonstration models were carefully inspected and the Scotts™ EasyGreen Model ER-2A in disassembled form was chosen. The reasons for the purchase of this particular machine instead of others were as follows:

- small size
- low cost
- neat and sturdy appearance
- easy to clean
- obvious care and attention given to detail design
- description on box suggested good control
- spreader matched lawn-care product that was to be used.

The box was opened and this author was so impressed by the way the parts were packed and presented for assembly that a full photographic record was made of the unpacking and assembly process. Figure 9.3 shows the cover page of the assembly instructions, and Figure 9.4 shows the assembly instructions (much reduced in scale). The assembly process was much simpler and quicker than expected, as the parts were designed to fit in only the correct way. For example, use of a hollow plastic pinion and shaft with an innovative hexagonal snap arrangement reduced assembly of the pinion gear, bearings, impeller, and agitator to a few seconds. The detail manufacturing drawing for the pinion gear and shaft is shown in Figure 9.5. The rate pointer illustrated in Assembly Step 4 (Figure 9.4) and in the operating instructions (Figure 9.6) was designed to snap on to the rate scale in a similar fashion.

Assembly, Use, and Care

Model ER-2A Rotary Spreader

Thank you
for purchasing a Scotts® EasyGreen® spreader. O.M. Scott & Sons introduced the first home lawn spreader in 1946 and has continued to offer superior quality that translates into more accurate application, a better-looking lawn, and a longer-lasting, more durable spreader. The lawn care industry's most extensive research, design, and testing have gone into making the EasyGreen the first rotary spreader with the accuracy of a Scotts drop spreader. With its durable rustproof parts, the EasyGreen is guaranteed to give you great results—now and for many years to come.

Before you begin
to assemble your EasyGreen spreader, please take a moment to read this publication. The assembly is really quite easy when you follow the simple directions we've provided. Just a few common hand tools are needed; they are shown to the right.

The only tools you'll need
to assemble this spreader:
- Adjustable or ⅜" open-end wrench
- Flat-blade screwdriver
- Wide-jaw or "Channel-Lock" pliers
- Hammer

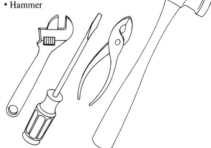

Use and care of your spreader
is also easy, especially with the helpful hints on the back of this publication. If you have any questions about the assembly or use of your EasyGreen, just call the toll-free number listed below. Our specially trained lawn consultants are waiting to help you.

Questions?
Call the Scotts toll-free Consumer Hotline: **1-800-543-TURF**

Figure 9.3. Scotts EasyGreen® Lawn Spreader. Courtesy of O.M. Scott & Sons Company

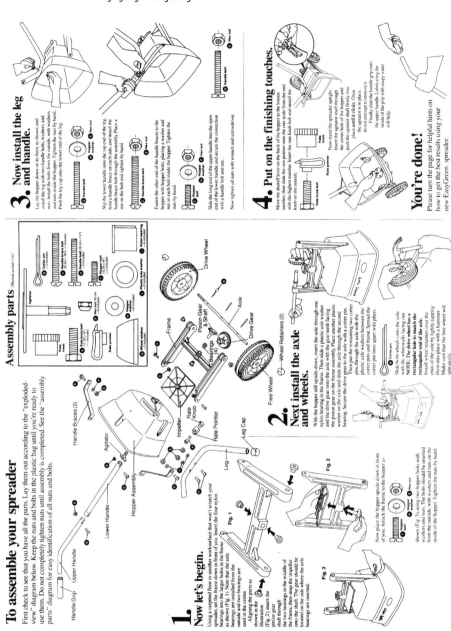

Figure 9.4. Lawn spreader parts and assembly instructions. Courtesy of O.M. Scott & Sons Company

Final Design: Detail Design for Manufacture   195

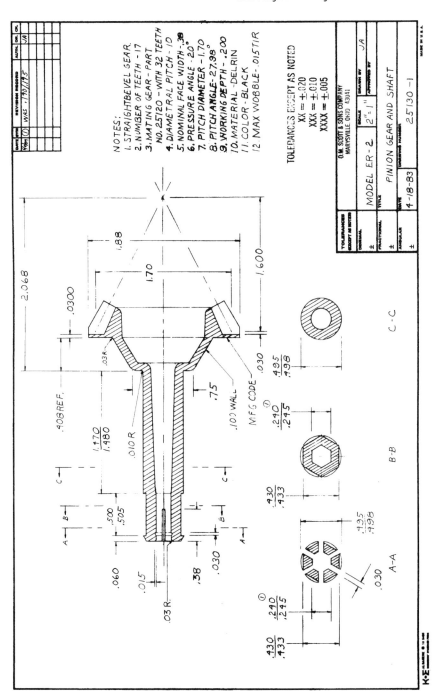

Figure 9.5. Detail drawing of pinion gear and shaft. Courtesy of O.M. Scott & Sons Company

To get the most from your new EasyGreen® spreader

Set the spreading rate before you begin

1. First check the back of your Scotts® product package for the proper rate setting number under the heading "Scotts EasyGreen." Some other lawn product manufacturers provide settings for Scotts spreaders, but using Scotts products will assure the best possible results. (If you already purchased a product which does not provide a setting for the Scotts EasyGreen spreader, you can use the trial-and-error method described in #5 below.)

2. Loosen the rate control knob and slide the pointer until it lines up with the number given for the spreader setting. Then tighten the knob.

3. Always fill the spreader on the driveway or sidewalk—not on the lawn. Before filling the spreader, move the shutoff lever as far left as possible to its closed position. Then pour the lawn product into the hopper.

4. When you're ready to begin, slide the shutoff lever to the right so it touches the pointer. Then begin walking at a normal pace. Product application automatically starts and stops when you do. Avoid tipping the spreader during use.
Always *push* the spreader when applying product. When pulled backward while open, the spreader may apply an excessive amount.
Move the shutoff lever to the closed position to stop the flow of product.

5. If the lawn product you purchased does not give the setting for your Scotts spreader, you can test for the proper setting.
Empty the number of pounds the manufacturer recommends for 1,000 square feet into the spreader. Adjust the spreader to a very low setting and spread product over a 1,000-square-foot area (50 by 20 feet, 33 by 30 feet, or equivalent). Adjust the setting upward after judging how much product has been applied to the test area. Remember: it is safer to under-apply with the test and increase setting later as needed. Some non-Scotts fertilizers can burn your lawn if over-applied.

Follow these application tips for guaranteed results

1. If your lawn is rectangular, apply product the longest direction. Apply two header strips across each end for a turning area. Steer smoothly around obstacles, keeping wheel about 2 feet from any area you do not wish to treat. CAUTION: Exercise care around ornamental plants, because weed controls can damage them.

2. If your lawn is an irregular shape, apply a header strip completely around it. Then apply product back and forth in the longest direction.

3. To avoid misses or "striping," make each pass about 30 inches from the previous one to partially overlap spreading patterns. You can do this by checking to see that particles are falling close to the previous tire tracks. For best results, do not apply product on a windy day.

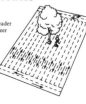

Spend a few minutes on maintenance after each use

1. Never leave product in the spreader. Pour leftover material back into the package and seal the package tightly.

2. After each use, wash thoroughly to remove all material clinging to the spreader. Hot water may be required at times to remove stubborn residue. Allow the spreader to dry thoroughly in the sun. Always store the spreader with it set wide open.

If you ever need replacement parts, just call 1-800-543-TURF

To order replacement parts, check the illustration closely to identify the name and number of the parts you need. Call our toll-free Consumer Hotline or write to Scotts Consumer Service, Marysville, OH 43041.
When ordering parts, prices will be furnished on request or parts will be shipped and billed at prevailing prices plus applicable taxes.

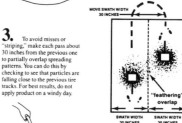

PARTS LIST—MODEL ER-2A SPREADER

Key No.	Description	Part No.	Key No.	Description	Part No.
1	Hopper assembly	25600	11	Drive wheel	25201
2	Frame	25100	12	Free wheel	25202
3	Axle	25111	13	Leg	25240
4	Rate pointer	25050	14	Upper handle	25250
5	Rate knob	25060	15	Lower handle	25260
6	Drive gear	25120	16	Leg cap	25170
7	Pinion gear and shaft	25130	17	Handle braces (2)	25270
8	Impeller	25140	18	Handle grip	25280
9	Agitator	25150	19	Wheel retainers (2)	15261
10	Bearings (4)	25160		Fastener pack	25501

Figure 9.6. Lawn spreader operating and maintenance instructions. Courtesy of O.M. Scott & Sons Company

9.9 Detail Design Checklist and Work Sheet

As for previous chapters, a checklist and work sheet were developed for use by the design manager in reviewing and assessing the work of the design team during this phase of the design process. The *Detail Design Checklist*, shown in Figure 9.7, provides a list of points covering aspects of detail design that need to be questioned by the design manager prior to manufacture. The *Detail Design Work Sheet*, shown in Figure 9.8, is used to record the status and action items, so that the design effort can be completed without missing out any important items.

Following the successful completion of a working prototype, Formway Furniture Ltd signed a licensing agreement with Knoll Inc., and the two companies collaborated to produce the detailed manufacturing information so that the Life chair could be manufactured on a mass-production scale. Although the detailed design information and manufacture of components was shared, the two companies had independent assembly facilities. Formway was to assemble and distribute chairs for the Australasian market, while Knoll supplied the USA and Europe. The work sheet in Figure 9.9 provides a worked example from the detail design of the Life chair. This work sheet considers the production of components for the supply to the global market (*i.e.* manufacture of all parts) and the assembly and distribution from Formway to supply to the Australasian market. Figure 9.9 is an assessment of the chair after the pre-production run. Chairs from this batch were to be used as demonstrators at the NeoCon trade fair.

The completed work sheet showed a good level of confidence in the *functional requirements* for the Life chair. Contributing factors were:

1. The *overall geometry* of the chair was considered to be good; however, some fine-tuning was required at a detailed level to correct tolerances on the seat pivot mechanism.
2. The *motion of parts* factor was good; however, the stiffness of some parts was different from the prototype due to the use of more cost-effective materials in the mass-production version.
3. The *forces involved* factor was marginal due to indications of poor performance on initial fatigue and creep tests. It was decided to revise materials selection for some parts to ensure adequate creep and fatigue strength.
4. The user input *energy needed* factor was good; however, it was influenced by the seat and back-mechanism tolerances; the long-term performance of these joints had yet to be proven.
5. The *materials* factor was marginal, owing to cost constraints prohibiting the use of some materials. This caused problems with structural properties, and new materials needed sourcing.
6. The *control system* influence was considered good. Special consideration had been placed on eliminating pinch points in the seat and back mechanism, resulting in an elegant solution.

DETAIL DESIGN CHECKLIST

REQUIREMENTS	CONTRIBUTING FACTORS	POINTS TO CONSIDER
FUNCTIONAL	Overall geometry Motion of parts Forces involved Energy needed Materials to be used Control system Information flow	Interference, assembly sequence, tolerances, surface finish Displacement, velocity, acceleration, position, fatigue, stiffness Weight of components, deflection, vibration, resonance, creep, flow, strength, residual stress Torque, speed, horsepower, power transmission Hardness, surface finish, friction, lubrication, replacement Button/switch design/layout, emergencies, safety, operation Assembly, operation, maintenance, safety
SAFETY	Operational Human Environmental	Modes of operation, abusive operation, maintenance Failure modes and effects analysis Specific issues related to design
QUALITY	Quality assurance Quality control Reliability	Certification, design and manufacture records Inspection and component testing, production documents Simulated tests/field tests/statistical analysis
MANUFACTURING	Production of components Purchase of components Assembly Transport	Manufactured as designed, revisions Inspection records Stacking, fit of parts, minimizing operation, ease of assembly Packaging, protection, storage, inventory control
TIMING	Design schedule Development schedule Production schedule Delivery schedule	Disruption caused by revisions Problem diagnosis, debugging procedure, testing materials Inventory control Acceptance criteria, commissioning
ECONOMIC	Marketing analysis Design costs Development costs Manufacturing costs Distribution costs	Customer reaction, user field tests, data collection Recording problems and solutions Cost of redesign Record of manufacturing problems/costs Record of packaging/distribution costs
ERGONOMIC	User needs Ergonomic design Cybernetic design	Functional performance, suggested improvements Understanding/use of instruction manual/controls/ease of use Reaction of machine to controls, feedback to user
ECOLOGICAL	Material selection Working fluid selection	Machining, assembly or operational problems found? Filling, spillage, leakage, maintenance, filtration
AESTHETIC	Customer appeal Fashion Future expectations	Surface finish, overall quality, colors, textures, consistency User reaction/comments/consumer reports Comment from field tests/commissions
LIFE-CYCLE	Distribution Operation Maintenance Disposal	Loading/unloading, labeling, transport mode Monitoring, feedback, returns, recalls Spare parts supply, spare parts inventory/tracking, service facilities Rebuild, remanufacture

Figure 9.7. Detail design checklist

Final Design: Detail Design for Manufacture

DETAIL DESIGN WORK SHEET   PROJECT: _____   DATE: _____

REQUIREMENTS	CONTRIBUTING FACTORS	CURRENT STATUS (Good / Marginal / Poor)	REQUIRED ACTION (Proceed / Revise / N/A)
FUNCTIONAL	Overall geometry Motion of parts Forces involved Energy needed Materials to be used Control system Information flow	☐ ☐ ☐ ☐ ☐	☐ ☐ ☐
SAFETY	Operational Human Environmental	☐ ☐ ☐ ☐ ☐	☐ ☐ ☐
QUALITY	Quality assurance Quality control Reliability	☐ ☐ ☐ ☐ ☐	☐ ☐ ☐
MANUFACTURING	Production of components Purchase of components Assembly Transport	☐ ☐ ☐ ☐ ☐	☐ ☐ ☐
TIMING	Design schedule Development schedule Production schedule Delivery schedule	☐ ☐ ☐ ☐ ☐	☐ ☐ ☐
ECONOMIC	Marketing analysis Design costs Development costs Manufacturing costs Distribution costs	☐ ☐ ☐ ☐ ☐	☐ ☐ ☐
ERGONOMIC	User needs Ergonomic design Cybernetic design	☐ ☐ ☐ ☐ ☐	☐ ☐ ☐
ECOLOGICAL	Material selection Working fluid selection	☐ ☐ ☐ ☐ ☐	☐ ☐ ☐
AESTHETIC	Customer appeal Fashion Future expectations	☐ ☐ ☐ ☐ ☐	☐ ☐ ☐
LIFE-CYCLE	Distribution Operation Maintenance Disposal	☐ ☐ ☐ ☐ ☐	☐ ☐ ☐

Figure 9.8. Detail design work sheet

DETAIL DESIGN WORK SHEET

PROJECT: **LIFE CHAIR** DATE: **JUNE 2002**

REQUIREMENTS	CONTRIBUTING FACTORS	CURRENT STATUS (Good / Marginal / Poor)	REQUIRED ACTION (Proceed / Revise / N/A)
FUNCTIONAL	Overall geometry	Marginal	Proceed
	Motion of parts	Good	Proceed
	Forces involved	Marginal	Marginal
	Energy needed	Good	N/A
	Materials to be used	Marginal	Revise
	Control system	Good	Proceed
	Information flow	Good	Proceed
SAFETY	Operational	Marginal	Revise
	Human	Marginal	Revise
	Environmental	Good	Proceed
QUALITY	Quality assurance	Poor	Revise
	Quality control	Poor	Revise
	Reliability	Poor	Revise
MANUFACTURING	Production of components	Marginal	Proceed
	Purchase of components	Poor	Revise
	Assembly	Marginal	Proceed
	Transport	Poor	Revise
TIMING	Design schedule	Poor	Proceed
	Development schedule	Marginal	Proceed
	Production schedule	Poor	Revise
	Delivery schedule	Poor	Proceed
ECONOMIC	Marketing analysis	Good	Proceed
	Design costs	Poor	Proceed
	Development costs	Good	Proceed
	Manufacturing costs	Poor	Revise
	Distribution costs	Poor	Proceed
ERGONOMIC	User needs	Good	Proceed
	Ergonomic design	Good	Proceed
	Cybernetic design	Good	Proceed
ECOLOGICAL	Material selection	Marginal	N/A
	Working fluid selection	Good	N/A
AESTHETIC	Customer appeal	Good	Proceed
	Fashion	Good	Proceed
	Future expectations	Good	Proceed
LIFE-CYCLE	Distribution	Poor	Revise
	Operation	Marginal	Revise
	Maintenance	Marginal	Revise
	Disposal	Good	Proceed

Figure 9.9. Example work sheet

7. The *information flow* factor was enhanced by publishing a comprehensive document of assembly instructions and by including general and technical information on a product Website.

Safety and compliance with safety standards needed to be verified for the final production version of the Life chair. Contributing factors were:

1. Although the prototype had been tested to international standards, extensive further testing was conducted to determine the safe *operational* life for the chair made from final materials and production processes. Maximum overload conditions and misuse scenarios such as "office-chair racing" were also considered.
2. Further work was in progress to correct *human* influences. The late change in materials made some safety standards difficult to obtain. Although many of Formway's competitors were not meeting all of the standards, Formway set compliance as a mandatory requirement.
3. *Environmental safety* was good because the chair delivered a good level of long-term support to the end user.

Factors contributing to *quality* included:

1. Formway had obtained certification for *quality assurance*; however, the team viewed this factor as marginal, because improvements were needed in keeping manufacturing records such as version controls.
2. *Quality control* systems were in place, but improvement was needed. The quality of components delivered by suppliers was inconsistent, which was partly due to poor control of production documentation.
3. *Reliability* needed to be proven for the production version. This would require hard and fair user trialing of the chair in its final working environment.

The Formway team was in unfamiliar territory in terms of high-volume *manufacturing*, and factors were assessed as follows:

1. The *production of components* factor was marginal to good: components were manufactured as intended; however, there were revision control issues that needed to be resolved.
2. The *purchase of components* factor was poor due to the lack of purchasing control; this needed to be revised by implementing an effective purchasing control system.
3. The initial product *assembly* sequences and operations were efficient; however, there were some minor incidences of product damage that had occurred during assembly. It was perceived that these incidences could be resolved by making minor detail design changes, such as including larger chamfers on holes to allow easier lead-in on some components.
4. *Transport* issues needed resolving. The packaging was efficient in terms of space saving; however, packaging needed redesign to reduce the likelihood of product damage.

Factors contributing to the *timing* requirements were:

1. The current status of the *design schedule* was poor due to time overruns when transferring the detailed manufacturing information from Formway in New Zealand to Knoll in the USA. This was attributed to the different CAD systems used by the two companies, and this had a strongly negative influence on the project.
2. The target for the *development schedule* was to produce a camera-ready product for the 2002 NeoCon trade fair in Chicago. The time scale for the product development was realistic; however, the time to get the collaborative working arrangement established took longer than expected.
3. *Production schedule* influence at Formway was poor. Knoll had a very good production schedule in place for their US manufacturing plant; however, at Formway, the inventory control system for the local market was inadequate.
4. The *delivery schedule* for the production version was too tight due to high customer demand, and this negatively influenced the project.

Drivers for *economic* requirements focused on reducing manufacturing costs to an acceptable level. Contributing factors were:

1. The *marketing analysis* of early production chairs involved "hard and fair" customer trials. Customer reaction was better than expected.
2. Formway and Knoll used different CAD software to produce their engineering drawings. At the detail design stage, Knoll re-modelled the Formway drawings using the different CAD system. This process was problematic and caused significant *design cost* overruns.
3. *Development costs* were positive due to lower than expected tooling costs. This had a positive influence on the project.
4. *Manufacturing costs* were strongly negative, as the team found that they needed to make unexpected cost reductions prior to the production ramp-up.
5. *Distribution costs* were higher than expected due to duty and currency differences. The high demand for the product also resulted in higher distribution costs because more urgent/expensive shipping was required.

Good *ergonomics* was critical to the long-term success of the chair. Contributing factors included:

1. Early customer trials showed that the chair performed in a reliable manner in terms of meeting *user needs*. Although the Formway team was happy with these early results they knew that the real test would be in long-term in-service performance of the chair and this had yet to be proven.
2. The *ergonomic and cybernetic design* requirements had been satisfied and in most cases exceeded. Some further work was identified to refine the lumbar control.

The *ecological* integrity of the prototype had been preserved in the production version. Some minor changes included more glass-filled product to achieve higher strength; however, the current status of *materials selection* was good.

User feedback showed a high level of *customer appeal* attributed to the *aesthetic* qualities of the first production version. From a *fashion* perspective the product was good because it had the "clean-green" minimalist aesthetic look. The product is also potentially adaptable to future office requirements, *e.g.* with allowance for a pallet for portable computing devices. *Future expectations* factors were good because issues had been resolved that allowed tool-less replacement of items such as seat covers.

Factors influencing the *life-cycle* requirements were:

1. *Distribution* had a strongly negative effect on the product life-cycle. This was due to product damage that occurred in shipping the product from the manufacturing facility. It was proposed that this issue be resolved by building closer relationships with the freight companies and by improvements in labeling and packaging.
2. *Operational* factors were found to be marginal due to time constraints, which did not adequately allow for in-service testing.
3. The *maintenance* factor was marginal because the spare parts supply and service facilities were not in place. The company did, however, have extensive experience in furniture maintenance, and they were committed to correcting in-service faults.
4. The *disposal* factor had a good influence on life-cycle requirements. All parts were recyclable and the use of composite materials was minimized. The chairs had a 10-year warranty, which increased the likelihood that they would remain in service for longer.

The completed work sheet shows that the level confidence in terms of the functional, safety, ergonomic, and aesthetic performance has been sustained in moving from the prototype through to producing the manufactured chair. As in the embodiment design work sheet (Figure 8.11), areas of weakness that remain are quality, timing, and economic factors. Furthermore, the detail design work sheet shows weaknesses in factors influencing the life-cycle requirements. Although the Formway team had a proven track record in the manufacture of products for the Australasian market, it was clear from the detail design work sheet that the design team needed to focus on producing the detailed production control documents required to make the product in much larger volumes. Converting the Formway CAD models over to suit the software used by Knoll was another stumbling block, and this had a most significant negative effect on timing requirements. From the results of this analysis, it is questionable as to whether the product should have been ramped up into full production before correcting these factors.

9.10 Tips for Management

- Detail design is of critical importance to a successful product.
- Carefully check sufficient details to ensure that design quality is acceptable.
- Pay careful attention to the interaction between shape, material, and manufacture.
- Check components for adequate strength and adequate stiffness.
- Check for problems with fatigue, residual stresses, flaws, tolerances, and surface finish.
- Check to see that the most appropriate material sections have been used.
- Check that heat treatments and surface treatments have been properly specified.
- Consider as many failure modes as possible, by formal analysis if necessary.
- Check that standard components have been used appropriately.
- Check that proper specifications have been called out for joining and fastening.
- Check to make sure that it is possible to assemble everything according to the drawings.
- Check the assembly instructions for accuracy, detail, and practicability.
- Ensure that adequate testing and commissioning procedures have been compiled.
- Ensure that all manufacturing and production documents are properly in order.

Chapter 10

Users and Customers: Design Feedback

10.1 Expectations
10.2 Use and Abuse
10.3 Maintenance
10.4 Litigation
10.5 Design Quality Assessment Work Sheet
10.6 Tips for Management

10.1 Expectations

Well-designed products tend to be readily accepted and absorbed into general usage, setting the standard until superseded by improved models or a completely new development. The expectations of customers and users change with time, not always in a predictable fashion. Once upon a time it was accepted practice to use a starting handle to hand crank one's car, and many a sore thumb resulted from hooking it over the handle in the wrong way. In 1912, Cadillac led the way in introducing starter motors. These soon became the norm, and hand-cranking became an annoyance one put up with to get going when the battery "went flat," which it did with monotonous regularity. Now the expectation is that the car should start at the turn of a key, thousands of times, year after year, without any attention whatsoever. Anyone designing a product must, as a minimum, meet current user expectations or the product will not sell, and the trick is also to try and work out how user expectations will change in the future so as to meet those better than the competition (Cagan and Vogel, 2002). For example, the concern for product quality, product safety, and environmental issues is likely to increase steadily, so design engineers must continually improve their knowledge and skills in these areas.

It is interesting that user expectations do not always *increase* and may change in unexpected ways. For example, mail-order catalogs for "high-tech" luxury items such as the motorized tie rack found that their products started to become used as examples of the wasteful "throwaway society" having complete disregard for the environment. The emphasis in the catalogs soon changed, and so did many of the products. Alongside the motorized tie rack appeared a home composting unit made from recycled plastic, and posters of endangered wildlife on recycled paper. Expectations also vary considerably from country to country, and products may have to be designed to adapt to this. For example, in the USA, all electrical appliances have been sold with a plug molded onto the electrical cord for decades, whereas in the UK, up until 1992, appliances were sold with three bare wires poking out of the cord. The user was expected to find a plug from somewhere else and wire it up (often wrongly) before the appliance could be used. It is reputed that one company in the USA had to set up a small production line for cutting plugs off cords, to meet the UK requirements!

10.2 Use and Abuse

Design engineers can no longer assume that only reasonable people will be using their products. Plaintiff's lawyers have seen to it that products must be designed to cater to the most extreme use of products, almost to the point where it has become ridiculous. Is it really the designer's problem if someone gets injured while using a rotary lawn mower to cut a hedge? Of course, many products were, and perhaps still are, inherently hazardous, but the burden currently put on design engineers to make a product "safe" no matter how it is used can severely limit the application of innovative ideas or the development of otherwise useful products. In the USA, one single serious product liability lawsuit can put a manufacturing company out of business. It is not only the cost, publicity, and aggravation surrounding the lawsuit itself, but also the destruction of the creative spirit and the retrenching to a position of such caution with regard to design of new products that the competitive edge is lost before any design starts.

Of course safety is important, and great improvements have been made through standards and regulations to ensure that products meet basic safety requirements before they are sold to the public. The problem is that on some issues, such as warnings, there has been overkill, whereas on others there are fundamental safety problems that have not been addressed adequately. A hammer, for example, is designed primarily to hit things with. Hitting things is a risky business, but very useful at times. If we did not have hammers then we would use something else to hit things with. In all honesty, can we really blame the hammer manufacturer if we hit our thumb instead of the nail? On the other hand, vertically opening garage doors counterbalanced with springs have a fundamental safety problem. When the door is closed the springs are tensioned, ready to injure you if they break or let loose. When the door is open it is ready to fall on your head if anything breaks or lets loose. The problem is that the door seems quite innocuous. It takes only a small force to move it up and down, and there is no "open and obvious" hazard as there is when hitting things with a hammer. The garage door system has a hidden hazard, in the sense that it is never in a "zero mechanical state" (Barnett and Switalski, 1988) while operational. There is a large amount of potential energy locked up in the system at all times, and this regularly causes mischief. A personal friend has had a narrow escape from injury several times when springs let loose on garage doors. One time he was up a ladder, which was destroyed beneath him by a flailing spring. Another time the spring completely disappeared and was not found for several years. When there is no injury such "near misses" go unreported, and if statistics are based only on reported incidents involving injury or death then the true extent of the hazard is not apparent. This is not to say we should eliminate such garage doors. They are generally a boon and a great improvement over their clumsy forerunners. However, it is a design area where safety needs to be considered at a deeper philosophical level than current practice.

Example: Hearing Protectors

Eschenbrenner v. Willson Safety Products, Arkansas, Independence County Circuit Court, No. CIV-86-28, June 2, 1989.

Eschenbrenner, 23, was wearing a set of Willson Model 358A noise-reduction headphones when a co-worker pulled the ear cups away from his head and let them go. The cups rotated so that their hard plastic sides snapped against Eschenbrenner's face and jaw. He suffered temporomandibular joint dysfunction, resulting in chronic pain. He is barely able to open his mouth and requires a liquid diet. A factory worker who had earned approximately US$16 000 annually, Eschenbrenner was able to return to work for a short time as a restaurant manager. However, he became disabled when activity aggravated his condition.

Eschenbrenner sued Willson, the manufacturer of the noise protectors, alleging defective design, in that the headphones should have had a brake mechanism. This would have prevented the ear cups from swivelling around so that their hard side faced the wearer's head. The plaintiff presented evidence that such a modification would have cost only 50 cents per unit.

The jury awarded US$590 513.

10.3 Maintenance

Another area of design where there have been dramatic changes over the years is with regard to the maintenance of products in service. It was not long ago when our cars had grease nipples sprouting from every joint and they had to be "taken in to be greased" or "lubed" every so many miles. In fact, many service facilities still use such terminology, and perhaps charge naive customers for a "greasing" they do not get because there is no longer anything to grease. New materials and designs have reduced routine car maintenance requirements to an extremely low level. However, this introduces another problem. The less maintenance that something needs and the longer it runs without trouble, the less attention it is likely to get from the user. Then, all of a sudden, the one item that should have been changed or checked on a regular basis fails in service and the design is called into question on the basis of terms such as *foreseeable misuse*. It is often possible to overcome this type of problem once it has been identified and reported back to the design engineers. For example, car brakes now last a long time, are self-adjusting, and require almost no maintenance. Yet if the bonded pads wear out on the front disks, then brakes that the driver has come to rely on may one day not be able to stop the car quickly enough to prevent an accident. An attorney friend came in one day complaining that his brakes were making the most frightful squealing noises all of a sudden. What on earth could be the matter? No, the pads had not been checked recently. And yes, that was the problem. The cunning design engineer had inserted metal "squealers" into the pads to alert the driver that the brakes needed checking. Not only that, but the noise is also designed to be so horrible that it forces the driver to do something about it immediately.

10.4 Litigation

When designing a "one-off" or "special-purpose" piece of equipment, such as our example of the gasifier test rig referred to previously, then it is likely that the users are known to the design team and have helped in the design. They are likely to be trained in the use of the equipment and if there are difficulties in service the designers can be called in to advise or modify something. The probability of an accident is low, although the consequences of an accident may be high. On the other hand, when a product is designed for the mass market, with users from all walks of life, in many different environments, the probability of an accident happening somewhere, sometime, is high, although the consequences may be low. In either case the question of liability cannot be ignored, but in the case of the mass-produced product it is almost certain that there will be product liability lawsuits to resolve, even with the most innocuous products. If there are a lot of people involved in the manufacture and distribution of the product then it is also likely that there will be other types of lawsuit as well. The design manager can no longer afford to avoid the issues in the hope of being lucky, or that the product is considered so good that nobody would ever attack it in court. The approach has to be one of minimizing the risk (in financial terms) to the company or its insurance carrier. Minimizing the probability of an accident happening by safety in design is only one aspect of minimizing the risk. There is a lot that the design manager can do to make sure that, if a lawsuit is filed against the company, the costs of the litigation process and the case settlement costs are also minimized.

Having a clearly defined, visible, and systematic approach to the design process is a move in the right direction. Design records, such as notebooks, should be carefully organized and kept, decisions logged, and the implications of later modifications thoroughly reviewed from the legal point of view. If the new version of the machine is advertised as being "safer" than the old model, does this imply that all the old ones in use are unsafe? It is essential that a product meets mandatory performance, safety, and other requirements, and to minimize risk it is generally important that products meet all applicable voluntary standards as well. To say that one did not know they existed undermines the credibility of the design team, and not being aware of changes to existing standards is damaging to say the least. Merely meeting the standards may not be a strong defense either. In many cases this is just a starting point for the expensive discussions that follow. It is important for the design manager to consult with corporate attorneys and understand how to develop a strong defense posture from the start.

Even the way the design team presents itself physically in the defense of its product has a strong bearing on the outcome of a case, especially if it goes all the way to a jury trial. Design engineers are trained to search for information, jot things down, ask questions, and try out new ideas. This approach is disastrous in court. Every word is recorded, each blink of the eye is observed, every comment is dissected, and attorneys enjoy needling during cross-examination. How does one defend that nifty little rail designed behind the seat on a railway track laying machine when the jury hears it described as an "iron bar pressing on the necks of the American Worker"? Companies manufacturing products such as nailing guns, where the chances of an accident are higher than normal, are now tending to give design teams specific training in how to present their work and handle product liability claims against the product.

Example: The Bicycle

WOMAN PARALYZED IN FALL FROM GIVEAWAY BIKE RECEIVES $7 MILLION IN LUMP-SUM SETTLEMENT

MADISON, Wis.—(By a BNA Special Correspondent)-A Wisconsin woman paralyzed after falling from a bicycle she got in a retail promotion received a $7 million lump-sum settlement from the retailer, importer, and manufacturer (*Klomberg v. American TV and Appliance*. Wis CircCt MilwaukeeCnty, No. 89-CV-011310, settlement approved 1/30/91).

Attorneys said a trust established for Nola J. Klomberg, 44, of Sussex, Wis., who was totally paralyzed from a brain injury in the fall, will receive £3.7 million of the settlement. Her husband, James Klomberg, will receive $750,000; a son, Kurt, $250,000; her attorneys $1.8 million; and $500,000 will go toward health care and pretrial costs.

According to her complaint, Klomberg fell off the Firenze bicycle in September 1988, after the front fork wobbled uncontrollably. Her son got the bike in 1986 as part of a "Get a Bike" promotion run by American TV and Appliance, where the family bought an entertainment center.

Retailer Paid $5.5 million

According to plaintiffs' attorney Robert L. Habush, Pacific Cycles Inc. of Taipei, Taiwan, made the bicycle to order for distributor Diversified Investments Corp. of Madison, Wis., and American TV of Madison, Wis.

Pacific paid $500,000 in the settlement. Diversified paid $1 million, and American paid the rest, the manufacturer's attorney said.

The importer and retail seller of the bicycle were held to a higher standard of care than usual because they put a label on the bicycle reading "Competition High-Tension Steel" that promoted it as competition caliber, Habush told BNA Feb. 13.

"After it got into this country, Diversified and American decided to put a label on the bicycle which attempted to portray it as a competition cycle with high-tension steel." Habush said. "That was an independent act besides the manufacturing defects."

According to the manufacturer's attorney, John M. Swietlik, "There is no such thing as high-tension steel." The correct term would be high-tensile steel for competition bikes, he said.

'Potentially Monstrous' Verdict Feared

American agreed to pay most of the settlement because the insurance policies of the manufacturer and importer were limited and American faced a "potentially monstrous" verdict, its attorney, Don Carlson, said. Under Wisconsin strict liability law, a retailer is held to the same standard as a manufacturer, he said.

"American insisted that the bike meet appropriate standards as part of its specifications," Carlson said. "If the bike had met the standards, it would not have been any problem."

Carlson said he could not discuss American's actions, including when the bike was labeled, due to potential legal exposure.

Carlson is with the Milwaukee law firm Riordan Crivello Carlson Mentkowski & Steeves. Habush is with Habush Habush & Davis, in Milwaukee. Swietlik is with Kasdorf, Lewis & Swietlik, also in Milwaukee.

Reprinted with permission from *Product Safety and Liability Reporter*, Vol. 19, No. 8, p. 189 (Feb. 22, 1991). Copyright 1991 by the Bureau of National Affairs, Inc. (800–372–1033).

10.5 Design Quality Assessment Work Sheet

At this stage, when the design work is complete and when the product is ready for manufacture, it is important to conduct an overall review of the design to make sure that nothing obvious has been overlooked during the design work. The *Design Quality Assessment Work Sheet* shown in Figure 10.1 poses a set of simple questions for the design manager to ask of the design team on all phases of the design process. Filling out the work sheet is a subjective exercise, but one that enables the design manager to record what level of performance is to be expected from the design and to highlight any possible weaknesses. An example quality assessment work sheet for the Life chair is shown in Figure 10.2. The assessments of the phases of task clarification, conceptual design, and embodiment design indicate a high level of design acceptability. The task clarification phase ensured that the product met the user needs and promoted stability in the market. The conceptual design phase resulted in the inclusion of novel design features in the final solution. The outcome of the embodiment phase was an elegant layout, which was demonstrated in prototype evaluations.

The assessment work sheet for the Life chair shows that the majority of factors influencing the detail design were marginal. The shape–material–manufacture interactions were hampered by late changes in materials to meet unexpected cost constraints. The structural properties were marginal due to the loss of "design intent" (*e.g.* changes in details such as fillet radii geometry) when changing drawings from one CAD format to another. While the testing and commissioning procedures were adequate in identifying the structural properties, they failed to identify other problems such as squeaking in dry-mechanism joints. The in-service use of the first few production chairs was problematic due to minor, but annoying, mechanism noise and other minor issues. As a consequence of this, Formway stopped production and completed a full review of the design detail phase. These detail issues were resolved and problems with products already in service were rectified. The hatched "check boxes" in the detail section of Figure 10.2 show the detail design improvements. The structural changes were made after the initial production run, and the other changes were made after the detail design review.

The Life chair case study is in marked contrast to many other design projects we have analyzed. For example, consider Figure 10.3. This shows a similar type of assessment for the aft field joint used in the solid rocket booster for the Space Shuttle Challenger (Hales, 1987, 1989), based on the Report of the Presidential Commission (1986). It is interesting to note that in this case the work sheet exhibits a general trend from high design acceptability at the beginning of the design process to low acceptability at the end. Perhaps the design weaknesses would have been handled differently if the managers had asked this simple set of questions as the project progressed.

10.6 Tips for Management

- Treat product liability seriously.
- Use a systematic approach to engineering design.
- Keep clean and full design records.
- Meet applicable regulations, standards, and codes.
- Carry out an overall design review prior to manufacture.
- Carefully attend to any design weaknesses identified.
- Educate design staff with regard to product liability lawsuits.

DESIGN QUALITY ASSESSMENT WORK SHEET

PROJECT:	DESIGN ACCEPTABILITY		
	High	Marginal	Low
TASK CLARIFICATION: 1. Design problem clearly defined? 2. Agreed design specification? 3. Specification circulated to all involved?	☐ ☐ ☐	☐ ☐ ☐	☐ ☐ ☐
CONCEPTUAL DESIGN I - CONCEPT GENERATION 1. Problem abstracted? 2. Broken into sub-functions? 3. Several concepts produced? 4. Many working principles considered? 5. Principles suitably combined?	☐ ☐ ☐ ☐ ☐	☐ ☐ ☐ ☐ ☐	☐ ☐ ☐ ☐ ☐
CONCEPTUAL DESIGN II - SELECTION AND EVALUATION 1. Concept variants firmed up? 2. Concept variants evaluated: Technical? Economics? 3. Concept weak spots identified? 4. Cost estimates developed? 5. Concept formally presented for approval?	☐ ☐ ☐ ☐ ☐ ☐	☐ ☐ ☐ ☐ ☐ ☐	☐ ☐ ☐ ☐ ☐ ☐

EMBODIMENT DESIGN I - OVERALL LAYOUT
1. Design simple?
2. Design function clear?
3. Design form clear?
4. Safety: Safe-life design?
    Fail-safe design?
    Redundancy built in?
    Protection built in?
    Warnings provided?
5. Primary checks: Function OK?
    Economics OK?
    Safety OK?
    Ergonomics OK?
6. Secondary checks: Production OK?
    Quality assurance OK?
    Assembly OK?
    Transport OK?
    Operation OK?
    Maintenance OK?
    Costs OK?
    Schedule OK?
	☐
	☐
	☐
	☐
	☐
	☐
	☐
	☐
	☐
	☐
	☐
	☐
	☐
	☐
	☐
	☐
	☐
	☐
	☐
	☐
		☐
		☐
		☐
		☐
		☐
		☐
		☐
		☐
		☐
		☐
		☐
		☐
		☐
		☐
		☐
		☐
		☐
		☐
		☐
		☐
			☐
			☐
			☐
			☐
			☐
			☐
			☐
			☐
			☐
			☐
			☐
			☐
			☐
			☐
			☐
			☐
			☐
			☐
			☐
			☐

EMBODIMENT DESIGN II - DETAIL LAYOUT
1. Force transmission paths: Flowlines OK?
    Deformation OK?
    Secondary forces a problem?
2. Appropriate division of tasks?
3. Self-help used: Self-reinforcing?
    Self-balancing?
    Self-protecting?
    Self-damaging?
4. Design stable?
5. Calculations appropriate, adequate, correct and checked?
6. Materials selected and used appropriately?
7. Applicable standards and codes met?
8. Bought out components selected and used appropriately?
9. Engineering drawings professionally completed and updated?
	☐
	☐
	☐
	☐
	☐
	☐
	☐
	☐
	☐
	☐
	☐
	☐
	☐
	☐
		☐
		☐
		☐
		☐
		☐
		☐
		☐
		☐
		☐
		☐
		☐
		☐
		☐
		☐
			☐
			☐
			☐
			☐
			☐
			☐
			☐
			☐
			☐
			☐
			☐
			☐
			☐
			☐

DETAIL DESIGN - COMPONENTS AND ASSEMBLY 1. Shape, material and manufacture interactions OK? 2. Strength, stiffness, fatigue, creep ... OK? 3. Residual stresses, flaws, corrosion allowance ... OK? 4. Tolerances, surface finish, dimensional stability ... OK? 5. Easy to assemble components without ambiguity? 6. Testing and commissioning procedures adequate? 7. Production and certification documents in order?	☐ ☐ ☐ ☐ ☐ ☐ ☐	☐ ☐ ☐ ☐ ☐ ☐ ☐	☐ ☐ ☐ ☐ ☐ ☐ ☐

Figure 10.1. Design quality assessment work sheet

DESIGN QUALITY ASSESSMENT WORK SHEET

PROJECT: *LIFE CHAIR* — DESIGN ACCEPTABILITY (High / Marginal / Low)

TASK CLARIFICATION:
1. Design problem clearly defined? — High
2. Agreed design specification? — High
3. Specification circulated to all involved? — High

CONCEPTUAL DESIGN I - CONCEPT GENERATION
1. Problem abstracted? — High
2. Broken into sub-functions? — High
3. Several concepts produced? — High
4. Many working principles considered? — High
5. Principles suitably combined? — High

CONCEPTUAL DESIGN II - SELECTION AND EVALUATION
1. Concept variants firmed up? — High
2. Concept variants evaluated: Technical? — High
   Economics? — High
3. Concept weak spots identified? — High
4. Cost estimates developed? — High
5. Concept formally presented for approval? — High

EMBODIMENT DESIGN I - OVERALL LAYOUT
1. Design simple? — High
2. Design function clear? — High
3. Design form clear? — High
4. Safety: Safe-life design? — High
   Fail-safe design? — High
   Redundancy built in? — High
   Protection built in? — High
   Warnings provided? — High
5. Primary checks: Function OK? — High
   Economics OK? — High
   Safety OK? — High
   Ergonomics OK? — High
6. Secondary checks: Production OK? — Marginal
   Quality assurance OK? — Marginal
   Assembly OK? — High
   Transport OK? — Marginal
   Operation OK? — High
   Maintenance OK? — High
   Costs OK? — High
   Schedule OK? — High

EMBODIMENT DESIGN II - DETAIL LAYOUT
1. Force transmission paths: Flowlines OK? — High
   Deformation OK? — High
   Secondary forces a problem? — High
2. Appropriate division of tasks? — High
3. Self-help used: Self-reinforcing? — High
   Self-balancing? — High
   Self-protecting? — High
   Self-damaging? — High
4. Design stable? — High
5. Calculations appropriate, adequate, correct and checked? — High
6. Materials selected and used appropriately? — High
7. Applicable standards and codes met? — High
8. Bought out components selected and used appropriately? — High
9. Engineering drawings professionally completed and updated? — High

DETAIL DESIGN - COMPONENTS AND ASSEMBLY
1. Shape, material and manufacture interactions OK? — Marginal
2. Strength, stiffness, fatigue, creep ... OK? — Marginal
3. Residual stresses, flaws, corrosion allowance ... OK? — High
4. Tolerances, surface finish, dimensional stability ... OK? — Marginal
5. Easy to assemble components without ambiguity? — High
6. Testing and commissioning procedures adequate? — Marginal
7. Production and certification documents in order? — Marginal

Figure 10.2. Example of design quality assessment work sheet for the Life chair

DESIGN QUALITY ASSESSMENT WORK SHEET

PROJECT: *AFT FIELD JOINT*	DESIGN ACCEPTABILITY		
	High	Marginal	Low
TASK CLARIFICATION:			
1. Design problem clearly defined?	■	☐	☐
2. Agreed design specification?	■	☐	☐
3. Specification circulated to all involved?	☐	■	☐
CONCEPTUAL DESIGN I - CONCEPT GENERATION			
1. Problem abstracted?	☐	☐	■
2. Broken into sub-functions?	☐	■	☐
3. Several concepts produced?	☐	☐	■
4. Many working principles considered?	☐	■	☐
5. Principles suitably combined?	☐	■	☐
CONCEPTUAL DESIGN II - SELECTION AND EVALUATION			
1. Concept variants firmed up?	■	☐	☐
2. Concept variants evaluated: Technical?	☐	■	☐
Economics?	■	☐	☐
3. Concept weak spots identified?	☐	☐	■
4. Cost estimates developed?	■	☐	☐
5. Concept formally presented for approval?	☐	■	☐
EMBODIMENT DESIGN I - OVERALL LAYOUT			
1. Design simple?	☐	☐	■
2. Design function clear?	☐	■	☐
3. Design form clear?	☐	■	☐
4. Safety: Safe-life design?	☐	■	☐
Fail-safe design?	☐	☐	■
Redundancy built in?	☐	■	☐
Protection built in?	☐	☐	■
Warnings provided?	☐	☐	■
5. Primary checks: Function OK?	☐	■	☐
Economics OK?	☐	■	☐
Safety OK?	☐	☐	■
Ergonomics OK?	☐	■	☐
6. Secondary checks: Production OK?	☐	■	☐
Quality assurance OK?	■	☐	☐
Assembly OK?	☐	☐	■
Transport OK?	■	☐	☐
Operation OK?	☐	☐	■
Maintenance OK?	☐	■	☐
Costs OK?	☐	■	☐
Schedule OK?	☐	■	☐
EMBODIMENT DESIGN II - DETAIL LAYOUT			
1. Force transmission paths: Flowlines OK?	☐	☐	■
Deformation OK?	☐	☐	■
Secondary forces a problem?	☐	■	☐
2. Appropriate division of tasks?	☐	☐	■
3. Self-help used: Self-reinforcing?	☐	☐	■
Self-balancing?	■	☐	☐
Self-protecting?	■	☐	☐
Self-damaging?	☐	☐	■
4. Design stable?	☐	☐	■
5. Calculations appropriate, adequate, correct and checked?	☐	☐	■
6. Materials selected and used appropriately?	☐	■	☐
7. Applicable standards and codes met?	■	☐	☐
8. Bought out components selected and used appropriately?	☐	■	☐
9. Engineering drawings professionally completed and updated?	■	☐	☐
DETAIL DESIGN - COMPONENTS AND ASSEMBLY			
1. Shape, material and manufacture interactions OK?	☐	■	☐
2. Strength, stiffness, fatigue, creep ... OK?	☐	☐	■
3. Residual stresses, flaws, corrosion allowance ... OK?	■	☐	☐
4. Tolerances, surface finish, dimensional stability ... OK?	☐	☐	■
5. Easy to assemble components without ambiguity?	☐	☐	■
6. Testing and commissioning procedures adequate?	☐	☐	■
7. Production and certification documents in order?	■	☐	☐

Figure 10.3. Example of design quality assessment work sheet for the Space Shuttle Challenger

Chapter 11
Standards and Codes

11.1 General Issues
11.2 Basic Definitions
11.3 Safety Standards
11.4 Some Reference Articles on Safety Standards
11.5 Some Reference Articles on International Standards
11.6 ISO 9000 International Standards for Quality Management
11.7 National Standards for Engineering Design Management
11.8 Tips for Management
11.9 Contact Information and URLs for Standards and Codes

11.1 General Issues

In coming up with a design that meets user needs in the best way possible, tradeoffs must be made amongst the requirements of function, safety, timeliness, cost, ergonomics, the environment, and aesthetics. Established standards and codes help to provide the design engineer with a basis for making judgments such as "how safe is safe enough" in a professionally acceptable manner. The application, interpretation, and development of appropriate standards, codes, and certifications are issues of increasing importance to design engineers, especially with the more global approach to design and manufacture. Not only do the local requirements vary widely from place to place, but also so do user expectations and attitudes concerning the performance of products and equipment. Such factors can create unanticipated delays and additional design costs that are sufficient to jeopardize the future of a complete project, as shown by the example in Hales and Poczynok (2001), especially when combined with cultural and language misunderstandings. This chapter on standards and codes is included simply to highlight a few important issues and to provide a useful list of international contacts for basic information. The massive task of trying to assemble a coherent picture of what standards and codes exist in different countries, and how they all relate to each other, is something that really was not practicable until the Internet became a reality, but now it is possible to get a good overview by visiting the International Organization for Standardization (ISO) Website and those of the various organizations listed. For this reason the list has been updated to include the Website address for each contact, both within the text and electronically on the CD accompanying the book.

11.2 Basic Definitions

As a starting point for some discussion on standards and codes, consider the following dictionary definitions.

Standard:
[T]hing serving as a basis for comparison; document specifying (inter)nationally agreed properties for manufactured goods etc.   (Oxford)
   A degree of quality, level of achievement regarded as desirable and necessary for some purpose.   (Longman)

Code:
Systematic collection of statutes, body of laws so arranged to avoid overlapping; set of rules on any subject.   (Oxford)
   A collection of statutes, rules etc. methodically arranged.   (Longman)

   Thus, standards are more concerned with setting a level of performance, quality or safety by the definition of criteria, whereas codes are more concerned with ensuring a level of performance, quality or safety through adherence to a set of rules or guidelines. The variety of each is enormous; the requirements vary from area to area, and often there are inconsistencies that are complicated to resolve. It may be very difficult for the design engineer even to determine which standards or codes apply under particular circumstances, let alone to interpret the details of the "fine print." Some are in ISO units, others are in imperial units, some are international, others are national, others are regional, and others are local. Some are specific to a particular product, whereas others are more generic; some deal with the minute details of a material composition, whereas others deal with the testing of whole assemblies or lay down safety procedures. Even the terminology used varies from one document to another, and subtle differences in meaning can sometimes lead to expensive misunderstandings. From an engineering design perspective, most codes and standards are helpful in that they:

- Define rules or criteria for basic safety.
- Differentiate mandatory ("shall") from recommended ("should") requirements.
- Offer additional guidance and commentary.
- Encapsulate field experience.
- Reflect a consensus.
- Are continually updated.
- Allow for special cases.

However, it must be appreciated that they:

- Are not "cookbooks" for design.
- Lag contemporary events, such as failures and accidents.
- Become increasingly complex with time.

From the design manager's point of view, the important thing is to know which standards and codes apply to the particular product in any of the countries where it may be used, how the applicable documents may be obtained, and what the implications are of noncompliance. Is the standard mandatory by regulation or jurisdictional adoption, is it voluntary within the industry, or does it simply reflect "accepted professional engineering practice"? Is it an industry-wide standard or a specific company standard? Who set the criteria and on what basis? Does the standard apply to only one product or to a range of similar products? Whereas there are many standards specific to the design of particular products, so far there only a few that address the engineering design process in general, such as the German standards VDI 2221 and VDI 2223, and the British BS 7000 series. With regard to the overall issue of product quality, the series of ISO 9000 International Standards for Quality Management (ISO, 2000) is becoming accepted worldwide as a means to help improve the efficiency and effectiveness of company operations. The national standards institutes of some 140 countries are now members of ISO. Specific terminology used in the standard is explained on the ISO Website page: http://www.iso.ch/iso/en/iso9000-14000/basics/general/basics_5.html. *Certification* means the written assurance by an appropriate independent, external organization to the effect that it has audited the company's management system and verified that it conforms to the requirements specified in the standard. *Registration* means that the auditing organization has then recorded the certification in its client register. By this process the company's management system is certified and registered. Certification and registration is not an ISO 9000 requirement, but the independent audit may be needed for business reasons such as:

- contractual or regulatory requirement;
- market requirement or to meet customer preferences;
- risk management;
- staff motivation.

Accreditation means that the organization providing certification has been officially approved as competent to carry out certification in the company's business area by a national accreditation body. As stated within the ISO Website information: "In most countries, accreditation is a choice, not an obligation and the fact that a certification body is not accredited does not, by itself, mean that it is not a reputable organization. For example, a certification body operating nationally in a highly specific sector might enjoy such a good reputation that it does not feel there is any advantage for it to go to the expense of being accredited. That said, many certification bodies choose to seek accreditation, even when it is not compulsory, in order to be able to demonstrate an independent confirmation of their competence."

It is important for the certification of a company to be carried out by a reputable and accepted organization, and the ISO suggests the following guidelines to help in the selection process:

- Evaluate several certification bodies.
- Ensure adequate standard of auditing, noting that the cheapest may eventually prove the most expensive if unacceptable.
- Ensure certificate is recognized by company customers.
- Establish that the certification body has auditors with relevant experience.
- Check that the focus of the certification body is on performance rather than on conformity.
- Clarify whether or not the certification body has been accredited and, if so, by whom.

Despite all this, it has been known for a company to become "ISO 9000 registered" simply by demonstrating that it meets all the paperwork requirements, while in practice its design process remains dysfunctional and of unacceptably poor quality. Unless the design manager ensures that the engineering design process as actually carried out within then the company meets the intent of the standard wholeheartedly then the company may be operating under false pretences, with a high risk of an accident or design failure. In the event of liability claims the company is then in a decidedly vulnerable position.

11.3 Safety Standards

The most controversial standards are often those concerned with safety. They are of concern to the design manager as they strongly influence the design, operation, and maintenance of technical systems and products. When accidents occur in which the design of a product or technical system is alleged to have been a contributor, the question of whether or not the design met applicable safety standards is pivotal.

A safety standard is a document intended to specify components and practices that will result in predictable and acceptable levels of safety. The concept of what is safe needs more careful definition in the design context than in general usage, and its definition is dependent on a set of related ones, as shown by the following definitions adapted from Hebert and Uzgiris (1989):

- *Hazard* – A condition or situation exhibiting the potential for causing *harm*.
- *Harm* – An adverse effect that occurs in an *accident*.
- *Accident* – An undesirable event or failure that results in *harm*.
- *Risk* – A measure of the probability and severity of *harm*; the potential of a *hazard* to cause *harm*.
- *Safe* – A characterization of a machine, product, process, or practice whose attendant *risks* are judged to be acceptable.
- *Safety* – A state or condition wherein people and property are exposed to a level of *risk* that is judged to be acceptable.

- *Safety standard*   – A set of criteria or means for achieving a level of *risk* that is judged acceptable by the body formulating the *safety standard*.

In terms of these definitions, the task of the design engineer is to design a safe system by identifying the hazards and controlling the associated risks to within acceptable limits. The criteria for what is acceptable are set, in part, by safety standards. Although compliance with applicable safety standards is generally understood to be a necessary condition for safe design, it may not be a sufficient condition. The standard may not have kept pace with industry or new developments, and it may not address all the hazards involved. It is up to the design engineer to identify hazards, whether or not they are described in the standard, and to make sure that the issues are adequately addressed.

11.4   Some Reference Articles on Safety Standards

The following articles and papers were prepared by the staff of Triodyne Inc., Northbrook, IL 60062, USA.

Barnett, R.L. (1983). On safety codes and standards. *Triodyne Inc. Safety Brief* **2** (1), 1–5 (www.triodyne.com).
Dilich, M.A., Rudny, D.F. (1989). *Compliance with Safety Standards: A Necessary but not Sufficient Condition*. Paper ASME 89-DE-1. ASME International, New York.
Hamilton, B.A. (1983). Managing a standards collection in an engineering consulting firm. *Special Libraries* **74** (1), 28–33.
Hansen, C.A., Hebert, J.J., Dilich, M.A. (1989). *Standards Identification and Retrieval for the Design Engineer*. Paper ASME 89-DE-2. ASME International, New York.
Hebert, J.J., Uzgiris, S.C. (1989). *The Role of Safety Standards in the Design Process*. Paper ASME 89-DE-3. ASME International, New York.

11.5   Some Reference Articles on International Standards

EC (1986). *A Journey Through the EC: Information on the Member States and the Development of the European Community*. Office for Official Publications of the European Communities, Luxembourg.
Al-Khalaf, K.Y. (1991). The Saudi Arabian Standards Organization (SASO). *ASTM Standardization News* **19** (9), 48–51.
ANSI Battles EC Code (1991). *Tooling & Production* **57** (3), 20.
Breitenberg, M. (1991). *Questions and Answers on Quality, the ISO 9000 Standard Series, Quality System Registration, and Related Issues (NISTIR 4721)*. US Department of Commerce, National Institute of Standards and Technology, Gaithersburg, MD.
EC Testing and Certification Procedures under the Internal Market Program. US Department of Commerce, International Trade Administration, Washington, DC, 1 November, 1991.
European Community '92 Update, Business America. US Department of Commerce, International Trade Administration, Washington, DC, 25 February, 1991.

Lipin, O.F. (1991). Gosstandart International: the USSR national system for standardization, metrology, and product quality control. *ASTM Standardization News* **19** (9), 44–47.

Reihlen, H. (1991). Standardization and certification in Europe – 1992 and beyond. *ASTM Standardization News* **19** (6), 38–43.

Saunders, M. (1991). *ISO 9000 and Marketing in Europe: Should US Manufacturers be Concerned?* US Department of Commerce, National Institute of Standards and Technology, Gaithersburg, MD.

Toth, R.B. (1984). Putting the US standards system into focus with the world. *ASTM Standardization News* (December), 16–20.

US and EC Improve Market Access Over Testing and Certification. Europe Now. US Department of Commerce, International Trade Administration, Washington, DC, September 1991.

11.6 ISO 9000 International Standards for Quality Management

ISO 9000: 2000 Quality Management Systems – Fundamentals and Vocabulary.
ISO 9001: 2000 Quality Management Systems – Requirements.
ISO 9004: 2000 Quality Management Systems – Guidelines for Performance Improvements.
ISO 19011 Guidelines on Quality and/or Environmental Management Systems Auditing.
ISO 10005: 1995 Quality Management – Guidelines for Quality Plans.
ISO 10006: 1997 Quality Management – Guidelines to Quality in Project Management.
ISO 10007: 1995 Quality Management – Guidelines for Configuration Management.
ISO/DIS 10012 Quality Assurance Requirements for Measuring Equipment – Part 1: Metrological Confirmation System for Measuring Equipment.
ISO 10012-2: 1997 Quality Assurance for Measuring Equipment – Part 2: Guidelines for Control of Measurement of Processes.
ISO 10013: 1995 Guidelines for Developing Quality Manuals.
ISO/TR 10014: 1998 Guidelines for Managing the Economics of Quality.
ISO 10015: 1999 Quality Management – Guidelines for Training.
ISO/TS 16949: 1999 Quality Systems – Automotive Suppliers – Particular Requirements for the Application of ISO 9001: 1994.

11.7 National Standards for Engineering Design Management

11.7.1 Germany

VDI 2221: Systematic Approach to the Design of Technical Systems and Products, May 1993.
VDI 2223: Systematic embodiment design of technical products, January 2004.
VDI 2234: Basic Economical Information for Design Engineers, January 1990.

11.7.2 UK

BS 7000-1: 1999 Design Management Systems. Guide to Managing Innovation.
BS 7000-2: 1997 Design Management Systems. Guide to Managing the Design of Manufactured Products.
BS 7000-3: 1994 Design Management Systems. Guide to Managing Service Design.
BS 7000-4: 1996 Design Management Systems. Guide to Managing Design in Construction.
BS 7000-5: 2001 Design Management Systems. Guide to Managing Obsolescence.

BS 7000-10: 1995 Design Management Systems. Glossary of Terms Used in Design Management.

11.8 Tips for Management

- Find out which codes and standards apply to your products or equipment.
- Set up a library of codes and standards, including company standards.
- Keep codes and standards up to date.
- Test products according to relevant mandatory and voluntary standards.
- Keep careful records of all testing procedures and test reports.
- Become involved with developing codes and standards through membership of committees.

11.9 Contact Information and URLs for Standards and Codes

The following organizations make their catalogs of standards available through their Websites. They provide the design manager with a starting point for building up a personal library of information on standards relevant to a particular design situation. The outlook for standards is continually changing, and the best way to keep abreast of the latest requirements is by direct contact through Website, telephone, or fax machine. The contact addresses and Websites listed were correct at the time of going to press, but they are likely to change with time, so space has been left beside each one for notes and updating.

11.9.1 International

International Organization for Standardization (ISO)
1, rue de Varembé, Case postale 56
CH-1211 Geneva 20
Switzerland
+ 41 22 749 01 11
www.iso.org
(ISO Key-Word-in-Context (KWIC Index) of International Standards)

International Electrotechnical Commission
3, rue de Varembé
PO Box 131
CH-1211 Geneva 20
Switzerland
+ 41 22 919 02 11
www.iec.ch

11.9.2 Europe

European Committee for Electrotechnical Standardization (CENELEC)
35, Rue de Stassart
B-1050 Brussels
Belgium
+ 32 2 519 6871
www.cenelec.org
(CENELEC Catalogue)

Russian State Committee for Standards
Leninsky Prospekt 9
Moscow, B-49, 119991
Russian Federation
+ 095 236 03 00
www.gost.ru/sls/gost.nsf

European Committee for Standardization (CEN)
36 Rue de Stassart
B-1050 Brussels
Belgium
+ 32 2 550 08 11
www.cenorm.be

The following institutions are CEN national members:

Asociación Española de Normalización y Certificación (AENOR)
Génova, 6
28004 Madrid
Spain
+ 34 91 432 60 00
www.aenor.es

Association Française de Normalisation (AFNOR)
Avenue Francis de Pressensé 11
93571 Saint Denis La Plaine Cedex
France
+ 33 1 41 62 80 00
www.afnor.fr
(AFNOR Catalogue)

British Standards Institution (BSI)
389 Chiswick High Road
London W4 4AL
United Kingdom
+ 44 208 996 90 00
www.bsi-global.com
(BSI Standards Catalogue)

Czech Standards Institute (CSNI)
Biskupsky dvúr 5
110 02 Praha 1
Czech Republic
+ 420 2 21 802 100
www.csni.cz

Dansk Standard (DS)
Kollegievej 6
2920 Charlottenlund
Denmark
+ 45 39 96 61 01
www.ds.dk

Deutsches Institut für Normung e.V. (DIN)
Postfach
10772 Berlin
Germany
+ 49 30 26 01 0
www.din.de
(DIN Technical Indexes German Standards)

Ente Nazionale Italiano di Unificazione (UNI)
via Battistotti Sassi 11B
20133 Milano MI
Italy
+ 39 02 70 02 41
www.uni.com

Finnish Standards Association (SFS)
Maistraatinportti 2
00240 Helsinki
Finland
+ 358 9 149 93 31
www.sfs.fi

Hellenic Organization for Standardization (ELOT)
313, Acharnon Street
11145 Athens
Greece
+ 30 210 21 20 100
www.elot.gr

Hungarian Standards Institution (MSZT)
Üllői str. 25
1091 Budapest
Hungary
+ 36 1 456 68 00
www.mszt.hu

Icelandic Standards (IST)
Laugavegur 178
IS-105 Reykjavik
Iceland
+354 520 7150
www.stadlar.is

Institut Belge de Normalisation (IBN)
Avenue de la Brabançonne 29
1000 Bruxelles
Belgium
+ 32 2 738 01 05
www.ibn.be

Instituto Português da Qualidade (IPQ)
Rua António Gião, 2
P-2829-513 Caparica
Portugal
+ 351 21 294 81 00
www.ipq.pt

Malta Standards Authority (MSA)
Second Floor, Evans Building
Merchant Street
MT-Valetta VLT 03
Malta
+ 356 21 24 24 20
www.msa.org.mt

National Standards Authority of Ireland (NSAI)
Glasnevin
Dublin 9
Ireland
+ 353 1 8073800
www.nsai.ie

Nederlands Normalisatie-instituut (NEN)
PO Box 5059
2600 GB Delft,
Netherlands
+ 31 15 269 03 90
www.nen.nl

Norwegian Standards Association (NSF)
PO Box 353 Skøyen
N-0213 Oslo
Norway
+ 47 22 04 92 00
www.standard.no/nsf

Österreichisches Normungsinstitut (ON)
Heinestraße 38
1020 Wien
Austria
+ 43 1 213 00
www.on-norm.at

Service de l'Energie de l'Etat (SEE)
Organisme Luxembourgeois de Normalisation
BP 10
2010 Luxembourg
Luxembourg
+ 352 46 97 46 1
www.see.lu

Slovak Standards Institute (SUTN)
Karloveská 63
PO Box 246
SK-840 00 Bratislava
Slovakia
+ 421 2 60 29 44 74
www.sutn.gov.sk

Swedish Standards Institute (SIS)
Sankt Paulsgatan 6
SE-118 80 Stockholm
Sweden
+ 46 8 555 520 00
www.sis.se

Schweizerische Normen-Vereinigung (SNV)
Bürglistrasse 29
8400 Winterthur
Switzerland
+ 41 52 224 54 54
www.snv.ch

11.9.3 North America

Association of American Railroads (AAR)
50 F Street, N.W.
Washington, DC 20001-1564
USA
+1 (202) 639-2100
www.aar.org

American National Standards Institute (ANSI)
25 West 43rd Street, 4th Floor
New York, NY 10036
USA
+1 (212) 642-4900
www.ansi.org
(ANSI Catalog of American National Standards)

American Petroleum Institute (API)
1220 L Street, N.W.
Washington, DC 20005-4070
USA
+1 (202) 682-8000
www.api.org

American Society for Testing and Materials (ASTM)
100 Barr Harbor Drive
West Conshohocken, PA 19428-2959
USA
+1 (610) 832-9585
www.astm.org
(ASTM Publications Catalog)

American Society of Mechanical Engineers (ASME)
Three Park Avenue
New York, NY 10016-5990
USA
+1 (212) 591-7722
www.asme.org
(ASME Publications Reference Catalog)

Canadian Standards Association (CSA)
5060 Spectrum Way
Mississauga, Ontario
L4W 5N6
Canada
+1 (416) 747-4000
www.csa.ca
(CSA Services and Information Catalogue)

National Institute of Standards and Technology (NIST)
100 Bureau Drive, Stop 2150
Gaithersburg, MD 20899-2150
USA
+1 (301) 975-4040
www.nist.gov

Institute of Electrical and Electronics Engineers (IEEE)
3 Park Avenue, 17th Floor
New York, NY 10016-5997
USA
+1 (212) 419-7900
www.ieee.org

National Fire Protection Association (NFPA)
1 Batterymarch Park
Quincy, MA 02269
USA
+1 (617) 770-3000
www.nfpa.org
(National Fire Codes)

National Technical Information Service (NTIS)
5285 Port Royal Road
Springfield, VA 22161
USA
+1 (703) 605-6000
www.ntis.gov

Society of Automotive Engineers (SAE)
400 Commonwealth Drive
Warrendale, PA 15096-0001
USA
+1 (724) 776-4841
www.sae.org
(SAE Handbook)

Triodyne Inc.
666 Dundee Road, Suite 103
Northbrook, IL 60062-2732
USA
+1 (847) 677-4730
www.triodyne.com
(Safety briefs)

Standards Council of Canada (SCC)
270 Albert Street, Suite 200
Ottawa, Ontario
K1P 6N7
Canada
(613) 238-3222
www.scc.ca

Underwriters Laboratories (UL)
333 Pfingsten Road
Northbrook, IL 60062-2096
USA
+1 (847) 272-8800
www.ul.com
(UL Standard for Safety Catalog)

11.9.4 Pacific Rim

Japanese Standards Association
4-1-24 Akasaka Minato-ku
Tokyo 107-8440
Japan
+81 3 3583 8005
www.jsa.or.jp
(JIS Yearbook)

Standards Australia
286 Sussex Street
Sydney, NSW 2000
Australia
+ 61 2 8206 6000
www.standards.com.au

Standards New Zealand
155 The Terrace
Private Bag 2439
Wellington
New Zealand
+ 64 4 498 5990
www.standards.co.nz

HERA Information Centre
NZ Heavy Engineering Research Association
PO Box 76-134, Manukau City
Auckland
New Zealand
+ 64 9 262-2885
www.hera.org.nz
(Collected papers on quality assurance)

Chapter 12
Engineering Design Process: Review and Analysis

12.1 Summary
12.2 Forensic Analysis of Engineering Design Issues
12.3 Analysis of the Engineering Design Process

12.1 Summary

The successful management of engineering design projects requires an understanding of the context within which the project takes place, a professional and systematic approach to the guiding of design team activities, and an ability to monitor and assess the quality of design work as it is completed. This is a challenging and complex task involving the handling of diverse influencing factors, continuous team building, monitoring of design progress, and facilitating technical reviews.

The *checklists* and *work sheets* introduced in this book provide a simple means to help an engineering design manager assess the influences on a project at five levels of resolution, to monitor design progress during a project, and to maintain the quality of designs produced. At the same time they provide a standardized record of the project management issues in a Web-based format. This historical data, which up to now has been difficult to collect from projects in industry, may be used for planning the management of future projects and is important for building up the design knowledge base of a company over a period of time.

A systematic approach is encouraged in using the checklists and work sheets, starting with those to help identify important factors influencing the project and progressing through those addressing specific design issues up to the point of a major design quality review prior to manufacture. Where appropriate, guidelines are offered to help the design manager in assessing the quality of the design work being produced before any hardware has been made or tested. The whole approach involves no more than asking a specific set of questions during each phase of the engineering design process, based on some recommended procedures and guidelines for each phase. Although simple, this can provide a revealing and coherent assessment of a design at any point along the way.

The *context model* or map of the engineering design process described in Chapter 2 provides a structure for the checklist and work sheet approach. It is also intended as an aid to visualizing the project from different viewpoints at the five following *levels of resolution*:

- External environment level
- Market level
- Company level
- Project level
- Personal level.

For management purposes, the design process associated with a particular project up to the final testing and manufacturing stage is considered to pass through the following four main phases, once a proposal or brief has been accepted:

1. *Task clarification*
   Through task clarification activities, the problem is defined.
   Output is an agreed design specification.
2. *Conceptual design*
   Through conceptual design activities, alternatives are generated, selected, and evaluated.
   Output is an accepted concept.
3. *Embodiment design*
   Through embodiment design activities, the chosen concept is developed and proved.
   Output is a definitive layout.
4. *Detail design*
   Through detail design activities, every component is completely specified.
   Output is manufacturing information.

There are now well-accepted guidelines that address specific issues throughout the design process, and from a management point of view it is helpful to categorize them according to these four phases. The checklists and work sheets associated with each phase then provide a consistent and structured way of making sure that important issues and engineering details are addressed and reviewed progressively during the course of the project. By the end of detail design it is intended that the engineering design manager should feel confident that the project has been handled in a professional manner and that the work sheet records provide a summary of what was done, in a manner analogous to quality assurance records compiled during product manufacture.

When it comes to a final assessment of the design for acceptance purposes, it is often difficult to establish realistic criteria on which to make sound judgments. The system of work sheets helps to generate a solid foundation of quantitative and qualitative data for making such final assessments. This becomes extremely important if there are problems with the product in service at a later stage and the design engineer is called on to justify decisions made during the course of the design process. It is much more difficult and expensive to reconstruct a design history after the fact than to record it on simple data sheets as the design progresses. The question of whether or not a design meets applicable codes or standards often arises during the lifetime of a product and, again, if consistent records have been kept during the design process then the question is readily answered. If no records exist then there is an element of doubt, which is difficult to dispel no matter how acceptable the design is in other ways. Often, there will be a generic requirement for the designer to "meet all applicable codes and standards," but it has become increasingly difficult to determine what the "applicable" codes or standards actually are. Indeed, there may be conflicting requirements, where different organizations cover the same issues, but from different points of view. In order to avoid future accusations, the design manager is well advised to identify all applicable documents and to resolve any apparent anomalies by formal agreement. Chapter 11 is intended to help, by providing sources to contact for identifying codes and standards that apply in different circumstances. It is not necessarily sufficient merely to meet applicable codes and standards, but at least they offer a consensus starting point for assessing the quality of an engineering design.

12.2 Forensic Analysis of Engineering Design Issues

The creative and innovative nature of engineering design, resulting in a continuous stream of new products and equipment for human use worldwide, involves both technical and financial risk. Inevitably there will be disputes, failures, and accidents, some of which may rise to the level of a major catastrophe. When an engineering failure occurs and the excitement over "what broke" dies down, the hunt for who to blame and who is going to pay becomes a main focus. The spotlight often turns to the design itself, and occasionally to the design process that led to the design. The systematic approach to managing the engineering design process as presented in this book may also be used to help investigate what went wrong when a failure occurs or where there is a dispute over design process issues.

Many products and items of equipment come into being without a fanfare, perform well in service, and pass quietly into oblivion as new designs take over. However, when there is an accident or some kind of failure, the most mundane design can suddenly become the focus of intense scrutiny and bitter argument. The legal profession in the USA, for example, has established for itself a set of criteria for what it regards as a "design defect," as distinct from a "manufacturing defect" or a "warnings and instruction defect" (American Law Institute, 1998). These criteria, together with definitions regarding negligence, form the basis on which a design and the design process creating it will be judged in the event of a product liability or engineering failure lawsuit. It is important for those involved in the engineering design process to become familiar with the various legal perspectives and criteria against which they may be judged if things go wrong (Barnett, 1998). It is also important to understand the negative influence that such criteria can have on future innovation (Hales, 1999). There is a tendency for lawyers to see design in simplistic "black and white" terms, especially if this is beneficial to the outcome of their case, and the root cause of the problem is only of interest if it helps the lawyer to win. Forensic engineering, or the application of engineering knowledge to legal problems, is an established field with well-developed techniques for the gathering of evidence, analysis of failures, investigation of accidents, review of safety issues, and the presentation of conclusions in the form of "expert opinions." "Opinions" in the sense of an expert in a forensic engineering case are concise summary statements of conclusion "to a reasonable degree of engineering certainty" developed from review and analysis of all the available evidence at the time. They are likely to be challenged, word-by-word, not only during the course of the particular case, but also every time the issue arises in future cases. A set of opinions, therefore, represents a personal position, which needs substantial evidentiary support to survive. A forensic engineer with the assignment to find out "what happened," based on all the available evidence, needs an analytical approach that will enable professional opinions to be developed independently from anyone else's personal assessment.

12.3 Analysis of the Engineering Design Process

Although there are many ways for a design to have come into being, when it comes to a retrospective analysis of what actually happened during a project, it is helpful to have a systematic and structured way of mapping the evidence. The more the information can be sequenced and broken down into measurable components the more likely that objective conclusions will result and the better the chance there will be of developing defensible opinions. A good starting point is simply to take each of the four design process phases, task clarification, conceptual design, embodiment design and detail design, and see what evidence is available on how each was done. The sequence is not as important as the assessment of the activities and output associated with each phase. There must have been some kind of problem or design specification to start with, and there must have been some kind of concept from which a final design evolved. The concept must have been developed to a greater or lesser degree in order for the thing to be made at all, and every component must have been detailed at least to the point where the product or system could be manufactured. The available documentation on activities in each phase then provides information and data for analysis. It can also be compared against any testimonial evidence as to what took place during the design process. Often, it is found that there were strong external influencing factors impinging on the design process at different levels of resolution and at various points in time, which had a negative effect on the project outcome.

The same checklists used for reviewing issues during the course of a project may also be used to analyze the design of an existing product and the engineering design process that created it in the first place. Such a systematic approach to analyzing failures in engineering design can provide a powerful means of identifying the root cause correctly, defining it precisely, and presenting the results in a concise and understandable fashion. By use of timeline analysis, logic diagrams, review of activity records, and detailed inspection of documents such as design specifications and drawings, it is often possible to develop precise opinions, each with an adequate basis for defense through depositions and trial if necessary. For example, in one very large case, the deposition testimony of the design team members described how various concepts were generated and evaluated before a final concept was selected, but by analyzing the project time records it was found that it would not have been possible to carry out that amount of design work within the time recorded. The conclusion was that the design had been copied. This was further corroborated by the fact that spelling and dimensioning mistakes on the CAD detail drawings were found to be identical to mistakes made on the drawings of another company years before.

Example: Automatic Hot Melt Coater/Laminator Machine

To protect large underground steel pipes from corrosion it is common to wrap them spirally with a laminated tape, the impermeable outer plastic layer of which squeezes a viscous paint-like adhesive against the pipe surface. Until the adhesive is applied to the pipe it is sandwiched between the outer layer and a thinner plastic film "release layer." The tape is manufactured as a continuous web, in a machine which unrolls the plastic sheet materials, injects hot, melted adhesive between the two layers and rewinds the laminated web onto a cardboard core, similar to that used for paper-towel rolls only much larger. When a finished roll gets to about 1 M in diameter the web has to be cut and transferred to a new core, already prepared with diagonally wound sticky tape to anchor the cut end of the web and maintain tension. Continuous production of the laminated web requires automatic roll changes, while maintaining precise control of heat, speed, and tension. Although this is not easy, it can be achieved by installing an indexing turret rewind station for automatically cutting the web and laying it onto a new core.

A tape supply company ordered just such a custom-designed automatic machine from a company specializing in hot-melt injection systems, which in turn subcontracted the design and manufacture of the web transport system to an appropriate machine manufacturer. During

Continued

commissioning, numerous problems surfaced. In particular, a "hot wire" cut-off device had been fitted, with a taut heated wire that was supposed to sever the laminated web by rapidly melting through it during the roll change cycle. This was fitted to the same swinging arm assembly that also operated the "lay-on roll", used to press the end of the cut web against the sticky tape on the new core. Unfortunately, the hot wire cut-off failed to work except on the thinnest materials. Even when the machine purchaser replaced the whole device with a pneumatically operated knife blade it was so unreliable that operators had to stand by with utility knives on each side of the machine so as to finish cutting the web manually if necessary. For many of the tape products the operators simply disabled the entire swinging arm assembly to give themselves enough room for cutting the web manually during every roll change. Of course, disabling the swinging arm assembly also disabled the lay-on roll, which had always functioned satisfactorily but happened to be fitted to the same arm assembly. As the machine had been designed to operate automatically, without any provision for operator intervention, cutting the web was a hazardous operation. On each side of the machine a person had to perch with one foot up on a steel bar to gain enough height, then lean across to the center of the web so as to start cutting from the middle outwards in unison with the other person. What was even more difficult and hazardous was laying the cut end of the web onto the sticky tape of the core by hand. This meant trying to simulate the line contact force of the lay-on roll by a wiping motion of one's hand, instead of it being applied automatically along the full length of the roll as designed.

One day an assistant operator, having already cut the web, was attempting to press the free end against the core along its length when his right hand fingers caught on the sticky tape and were pulled into the rewinding web as it started its first wrap around the core. He was hauled over the top of the roll and dropped on to the floor, while his right arm was torn from his body as the trapped hand continued to be wound into the roll. Although he survived, his right arm was gone. During the ensuing lawsuit, which involved all the parties, there were a great many issues argued back and forth. For several years the defendants stuck to the theory that the plaintiff simply lost his balance while cutting the web and somehow caught his hand in the "nip point" underneath, where the web fed into the roll. However, the machine geometry and testimonial evidence clearly showed otherwise. The *first function* of cutting the web had already been completed before any problem occurred. It was during the *second function*, of laying and pressing the web onto the new core, when the accident happened. The immediate cause of the accident was the fact that the *two critical functions* of cut-off and lay-on had been *combined* on the same swinging arm *instead of being separated*, when it was known that the hot-wire cut-off was an unproven concept, installed without field testing. Had these two functions been separated and designed so as to operate independently from each other, as recommended by Pahl and Beitz (1984) in their embodiment design guidelines (see Section 8.3.2), disabling of the unusable cutting mechanism would not have affected the normal function of the automatic lay-on roll and this accident would not have happened.

The root cause of the accident was a *defective engineering design process*:

- Design specification — omission of requirement to provide for safe manual operation.
- Conceptual design — selection of hot wire cut-off concept without proving it would work.
- Embodiment design — combining critical functions for economy instead of separating them for reliability.
- Detail design — detailing and installing an unproven concept without contingency design.

There were also contributing factors from *negative influences on the design process*:

- Corporate structure — sale and restructuring of hot-melt injection company during project.
- Corporate systems — inadequate project communication during company reorganization.
- Customer expectations — unfulfilled key expectation of automatic machine operation.
- Customer involvement — attempts to rectify machine without sufficient design expertise.

All of which led to the following *consequences*:

- Late delivery of machine, with recriminations and lost production.
- Unacceptable machine performance for 3 years.
- Horrible accident to valued employee.
- 5 years of litigation regarding accident, settled 10 years after machine manufacture.
- US$2 million settlement cost on a US$1.4 million project.
- Bankruptcy of machine manufacturer.

In summary, the *activities of* the design team, the *influences on* the design team, and the consequent *output from* the design team are key issues, both in managing engineering design and in any forensic analysis of the design process. By applying what has been presented in this book it is possible to review design situations in a systematic, efficient, and effective way. The more precisely and confidently an analysis of the design process can be carried out and presented, the less likelihood of design failures in the first place and the more likely that any dispute over engineering design issues can be resolved at an early stage, thereby helping to reduce the enormous losses associated with accidents, failures and legal liability.

References

Alpert, S. (2003). *Design Reviews for Effective Product Development.* SAE International, Warrendale, MI. (Continuing Education Short Course – see http://www.sae.org).

American Law Institute (1998). *Restatement of the Law Third – Torts: Products Liability*, as adopted and promulgated by the American Law Institute at Washington, DC on May 20, 1997. American Law Institute Publishers, St Paul, MN (ISBN 0-314-23194-3).

ASI (1990). *Taguchi Methods: Selected Papers on Methodology and Applications.* American Supplier Institute, Dearborn, MI.

ASI/JSA (1990). *Taguchi Methods: Case Studies from the U.S.A. and Europe.* American Supplier Institute/Japanese Standards Association, Dearborn, MI.

ASM (2003). *ASM Handbooks Online.* Full 20-Volume ASM Handbook series available for Web. ASM International, Materials Park, OH.

Barnett, R.L. (1998). Design defect: doctrine of alternative design. *Safety Brief* 13 (4). Triodyne Inc., Northbrook, IL 60062, USA.

Barnett, R.L., Switalski, W.G. (1988). Principles of human safety. *Safety Brief* 5 (1). Triodyne Inc., Northbrook, IL 60062, USA. (ISSN 1041-9489)

Bazant, Z.P., Zhou, Y. (2002). Why did the World Trade Center collapse? Simple analysis. *Journal of Engineering Mechanics, Transactions ASCE.*

Belbin, R.M. (1981). *Management Teams – Why They Succeed or Fail.* Heinemann, London.

Birmingham, F.A. (1991). *Quality Improvement Guidelines.* Scott Fetzer Companies, Westlake, Ohio.

Cagan, J., Vogel, C.M. (2002). *Creating Breakthrough Products – Innovation from Product Planning to Program Approval.* Prentice Hall PTR, New Jersey.

Cebon, D. (2003). *Cambridge Engineering Selector, CES4: The CES Family of Software.* Granta Design Ltd, Cambridge, UK.

Clausing, D. (1994). *Total Quality Development: A Step-by-Step Guide to World-Class Concurrent Engineering.* ASME Press, ASME International, New York.

Darnell, H., Dale, M.W. (1982). Total project management – an integrated approach to the management of capital investment projects in industry. *Proceedings of the Institution of Mechanical Engineers*, 196 (36), 337–346.

Ealey, L.A. (1988). *Quality by Design: Taguchi Methods and U.S. Industry.* American Supplier Institute, Dearborn, MI.

Ehrlenspiel, K., Kiewert, A., Lindemann, U. (1998). *Kostengünstig Entwickeln und Konstruieren – Kostenmanagement bei der integrierten Produktentwicklung.* Springer-Verlag, Berlin.

FEMA (2002). *World Trade Center Building Performance Study: Data Collection, Preliminary Observations and Recommendations.* FEMA 403. Federal Emergency Management Agency, Washington, DC.

French, M.J. (1999). *Conceptual Design for Engineers* (3rd edn). Springer-Verlag, Berlin.

Giddens, P. (2003). *Micro-Hydro Turbines (Client – 2001 Design Project)*, Department of Mechanical Engineering, University of Canterbury, Christchurch, New Zealand (http://www.mech.canterbury.ac.nz/courses/Projects_01.htm).

Hajek, V.G. (1977). *Management of Engineering Projects.* McGraw-Hill, New York.

Hales, C. (1987). *Analysis of the Engineering Design Process in an Industrial Context*, PhD dissertation, University of Cambridge. Gants Hill Publications, Eastleigh, UK.

Hales, C. (1989) Analysis of an engineering design – The Space Shuttle Challenger. In: Samuel, A.E. (ed.), *Engineering Design and Manufacturing Management.* Elsevier, Amsterdam, chapter 9.

Hales, C. (1999). Legal threats to innovation in design. In: *Proceedings of ICED-99: WDK 26 (1) Communication and Cooperation of Practice and Science*, Technische Universität München, Garching, Germany, pp. 41–46.

Hales, C., Pattin, C. (2002). Design review for failure analysis and prevention. In: *ASM Handbook, Vol. 11: Failure Analysis and Prevention.* ASM International, Materials Park, OH, pp. 40–49.

Hales, C., Poczynok, P.J. (2001). From patent to product – the long hard road. In: *Proceedings of ICED-01: WDK 28, Design Applications in Industry and Education.* The Institution of Mechanical Engineers, London, pp. 99–106.

Hales, C., Howes, M.A.H., Bhattacharyya, S. (1981). Development of a laboratory test facility for high pressure erosion/corrosion evaluation of coal conversion structural materials. In: Hill VH, Black (eds), *The Properties and Performance of Materials in the Coal Gasification Environment.* American Society for Metals, Metals Park, OH, pp. 605–628.

Hales, C., Stevens, K.J., Daniel, P.L., Zamanzadeh, M., Owens, A.D. (2002). Boiler feedwater pipe failure by flow-assisted chelant corrosion. *Journal of Engineering Failure Analysis* **9**, 235–243.

Hampden-Turner, C., Trompenaars, A. (1993). *The Seven Cultures of Capitalism.* Doubleday, New York.

Hebert, J., Uzgiris, S.C. (1989). *The Role of Safety Standards in the Design Process.* ASME 89-DE-3. ASME International, New York.

Heilmann, J. (2001). Reinventing the wheel. *Time Magazine* 2 December.

Hughes Aircraft Company (1978). *R & D Productivity* (2nd edn). Hughes Aircraft Company, Culver City, CA.

Hunter, T.A. (1992). *Engineering Design for Safety.* McGraw-Hill, New York.

Intermediate Technology Development Group (2003). *ITDG Technical Briefs (On-line).* Intermediate Technology Development Group, Bourton Hall, Bourton-on-Dunsmore, Rugby, UK (www.itdg.org).

ISO (2000). *ISO 9000, International Standards for Quality Management.* International Organization for Standardization, Geneva.

Jones, J.C. (1970). *Design Methods – Seeds of Human Futures.* John Wiley & Sons, Chichester.

Lawrence, P., Lee, R. (1984). *Insight into Management.* Oxford University Press, Oxford.

Leech, D.J., Turner, B.T. (1985). *Engineering Design for Profit.* Ellis Horwood Engineering Science Series. John Wiley & Sons, Chichester.

Morrison, S.J. (1985). Quality management. *Proceedings of the Institution of Mechanical Engineers* **199** (B3), 153–159.

Newland, D.E., Cebon, D. (2002). Could the World Trade Center have been modified to prevent its collapse? *Journal of Engineering Mechanics, Transactions ASCE* **128** (7), 795–800.

Oakley, M. (ed.) (1990). *Design Management: A Handbook of Issues and Methods.* Basil Blackwell Ltd, Oxford, chapter 31.

Pahl, G., Beitz, W. (1984). Wallace, K.M. (ed.), *Engineering Design – A Systematic Approach.* The Design Council, London.

Pahl, G., Beitz, W. (1996). Wallace, K.M. (ed.), *Engineering Design – A Systematic Approach* (2nd edn translated by Wallace, K., Blessing, L., Bauert, F.). Springer-Verlag, Berlin.

Peters, T.J., Waterman, R.H. (1982). *In Search of Excellence.* Warner Books, New York.

Presidential Commission (1986). *Report of the Presidential Commission on the Space Shuttle Challenger Accident,* 6 June, Washington, DC.

Pugh, S. (1990). *Total Design: Integrated Methods for Successful Product Engineering.* Addison-Wesley, Wokingham.

Rodwell, C. (1971). Engineering design management. CME (Chartered Mechanical Engineer), Institution of Mechanical Engineer (July), 243.

Ryssina, V.N., Koroleva, G.N. (1984). Role structures and creative potential of working teams. *R & D Management* **14** (4), 233–237.

Schumacher, E.F. (1973). *Small is Beautiful.* Blond and Briggs Ltd, UK. (ABACUS edition published 1974, Sphere Books Ltd.)

Stetter, R., Ullman, D.G. (1996). Team-roles in mechanical design. In: *1996 ASME Design Engineering Technical Conferences.* ASME International, New York, paper 96-DETC/DTM1508.

The Technology Partnership (1988). *Product Development (Seminar Notes).* The Technology Partnership, Melbourn, Hertfordshare, UK.

Turner, B.T., Williams, M.R. (1983). *Management Handbook for Engineers and Technologists.* Business Books Ltd, London.

Ullman, D.G. (1992). *The Mechanical Design Process.* McGraw-Hill, New York.

Ulrich, K.T., Eppinger, S.D. (2004). *Product Design and Development* (3rd edn). McGraw-Hill, New York.

Wallace, K.M. (1984). Design for manufacture. Paper 2, Production Engineering Tripos, University Engineering Department, Cambridge (September).

Warby, D.J. (1984). Preparing the offer. *Proceedings of the Institution of Mechanical Engineers* **198B** (10).

Whybrew, K., Raine, J.K., Dallas, T.P., Erasmuson, L. (2002). A study of design management in the telecommunications industry. *Proceedings of the Institution of Mechanical Engineers* **216B**, 13–23.

Bibliography

Ackroyd, J. (1985). Thrust 2 – design of the world land speed record car. James Clayton Lecture. *Proceedings of the Institution of Mechanical Engineers* **199** (79).
Adamson, G. (ed.) (2003). *Industrial Strength Design – How Brooks Stevens Shaped your World*. The MIT Press, Cambridge, MA.
Alexander, M. (1985). Creative marketing and innovative consumer product design – some case studies. *Design Studies* **6** (1), 41.
Andersen, S. (1982). *Use of Methods for Need Assessment in Product Innovation – a Way of Improving the Utility of New Products*. University of Trondheim, Norwegian Institute of Technology, Norway.
Andreasen, M.M., Hein, L. (1987). *Integrated Product Development*. IFS (Publications) Ltd Bedford/Springer-Verlag, Berlin.
Antonsson, E.K., Cagan, J. (2001). *Formal Engineering Design Synthesis*. Cambridge University Press, Cambridge.
Arup, O. (1985). *The Arup Journal* **20** (1). Ove Arup's 90th Birthday Issue, Ove Arup Partnership, London.
Ashby, M.F. (1999). *Materials Selection in Mechanical Design* (2nd edn). Butterworth-Heinemann, Oxford.
Ashley, S. (1991). Underwriters Laboratories: safer products through tough tests. *Mechanical Engineering* (January), 63–65.
Becker, W.T., Shipley, R.J. (eds) (2002). *ASM Handbook, Vol. 11: Failure Analysis and Prevention*. ASM International, Materials Park, OH.
ASME Boiler and Pressure Vessel Code, Section VIII, Division 1. American Society of Mechanical Engineers (ASME International), Fairfield, NJ (published every 3 years).
Baker, G.J. (undated). *Quality Assurance in Engineering Design*. The Institution of Engineering Designers, Westbury, Wiltshire.
Barnett, R.L. (1998). Patents: restoring safety. *Safety Brief* **14** (1). Triodyne Inc., Northbrook, IL 60062, USA (ISSN 1041-9489).
Barnett, R.L. (1998). Reasonably foreseeable use. *Safety Brief* **14** (3). Triodyne Inc., Northbrook, IL 60062, USA (ISSN 1041-9489).
Baxter, M. (1996). *Product Design – Practical Methods for the Systematic Development of New Products*. Chapman & Hall, USA.
Blake, R., Mouton, J. (1964). *The Managerial Grid*. Gulf Publishing Co., Houston, TX.
Blockley, D.I., Robertson, C.I. (1983). An analysis of the characteristics of a good civil engineer. *Proceedings of the Institution of Civil Engineers, Part 2* **75** (March), 77–93.
Bohling, S. (ed.) (2002). The meaning of life. *PRODESiGN, Journal of the Designers Institute of New Zealand* (August/September), 24–28.
Bolster, C.F. (1984). Negotiating: a critical skill for technical managers. *Research Management* (November/December), 18–20.
Breitenberg, M. (1991). *Questions and Answers on Quality, the ISO 9000 Standard Series, Quality System Registration, and Related Issues*. NISTIR 4721. US Department of Commerce, National Institute of Standards and Technology, Gaithersburg, MD.
Bryant, D. (1982). Preparing a technical specification. CME (Chartered Mechanical Engineer), Institution of Mechanical Engineer (March), 45–49.
BS 5500 (1982). *Specification for Unfired Fusion Welded Pressure Vessels*. British Standards Institution, London.
BS 6046 (1984). *Use of Network Techniques in Project Management. Part 1: Guide to the Use of Management Planning, Review and Reporting Procedures*. British Standards Institution, London.
BS 7000-1 (1999). *Design Management Systems. Guide to Managing Innovation*. British Standards Institution, London.

BSI PD 6112 (1967). *Guide to the Preparation of Specifications (under revision)*. British Standards Institution, London.
BSI PD 6470 (1983). *The Management of Design for Economic Production*. British Standards Institution, London.
Carter, R., Martin, J., Mayblin, B., Munday, M. (1984). *Systems, Management and Change*. Harper & Row Ltd/The Open University, London.
Anon. (1996). *CE Mark: The New European Legislation for Products*. ASME Press, ASME International, New York.
Clarkson, P.J., Coleman, R., Keates, S., Lebbon, C. (2003). *Inclusive Design – Design for the Whole Population*. Springer-Verlag, Berlin.
Clements, S.D. (1985). Redesigning the engineer. *Proceedings of the Institution of Civil Engineers, Part 1* **78** (October), 1191–1202.
Coaker, J.W. (1980). *No Man is an Island (Analysis of Three Mile Island Incident)*. Paper 80-WA/Mgt-5. ASME International, New York.
Commerce Business Daily. (A daily list of US Government procurement invitations, contract awards, subcontracting leads, sales of surplus property and foreign business opportunities). Government Printing Office, Washington, DC.
Coover, H.W. (1984). Leadership – key to excellence in innovation. *Proceedings of IRI Spring Meeting, Medalist Address*. Industrial Research Institute, Boca Raton, FL.
Cross, N. (ed.) (1984). *Developments in Design Methodology*. John Wiley & Sons, Chichester.
Cross, N., Roy, R. (1975). *Design Methods Manual*. The Open University, Milton Keynes.
De Boer, S.J. (1989). *Decision Methods and Techniques in Methodical Engineering Design*. Academisch Boeken Centrum, De Lier, Netherlands.
Anon. (1980). *Defence Standard 05-67/1, Guide to Quality Assurance in Design*. HMSO, London.
Dieter, G.E. (1983). *Engineering Design – A Materials and Processing Approach*. McGraw-Hill, New York.
Elinski, D. (1985). Standards from the users perspective. *Standards Engineering* (March/April), 29–32.
Engineering Council. (1986). *Managing Design for Competitive Advantage*. The Engineering Council/The Design Council, London.
Es-Said, O.S. (2000). Scientific materials selection processes. In: Scutti, J. (ed.), *Proceedings of ASM/ASME Conference on Failure Prevention Through Education – Getting to the Root Cause*. ASM International, Materials Park, OH, pp. 70–80.
Evans, B. (1985). Japanese-style management, product design and corporate strategy. *Design Studies* **6** (January), 25–33.
Finkelstein, L., Finkelstein, A.C.W. (1983). Review of design methodology. *IEE Proceedings, Part A* **30** (4).
Flurscheim, C. (1977). *Engineering Design Interfaces*. Design Council, London.
Frankenberger, E., Badke-Schaub, P., Birkhofer, H. (eds) (1998). *Designers – The Key to Successful Product Development*. Springer-Verlag, London.
French, M.J. (1971). *Engineering Design – The Conceptual Stage*. Heinemann, London.
Gans, M. (1976). The A to Z of plant startup. *Chemical Engineering* (March 15), 72–82.
Gardiner, P., Rothwell, R. (1985). Tough customers: good designs. *Design Studies* **6** (1), 7–17.
Gatiss, J. (1981). The personality of the effective design engineer in manufacturing industry. In: Jacques, Powell (eds), *Design: Science: Method*. Westbury House, Guildford, pp. 176–178.
George, B. (1983). A contrast in style – Europe and the USA. *CME* (Chartered Mechanical Engineer), Institution of Mechanical Engineer (March), 47–48.
George, W.W. (1977). Task teams for rapid growth. *Harvard Business Review* (March–April), 71–80.
Glegg, G.L. (1969). *The Design of Design*. University Press, Cambridge.
Grant, C. (1985). *An Economics Primer*. Basil Blackwell Ltd, Oxford.
Gregory, S. (1983). Design for cost manufacture in West Germany. *Design Studies* **4** (4), 245–257.
Hales, C. (1985). Designer as chameleon. *Design Studies* **6** (2), 111–114.
Hales, C. (1990). Proposals, briefs and specifications. In: Oakley, M. (ed.), *Design Management: A Handbook of Issues and Methods*. Basil Blackwell Ltd, Oxford, chapter 31.
Hales, C. (2000). Ten critical factors in the design process. In: Scutti, J. (ed.), *Proceedings of ASM/ASME Conference on Failure Prevention Through Education – Getting to the Root Cause*. ASM International, Materials Park, OH, pp. 49–55.

Handy, C.B. (1981). *Understanding Organizations* (2nd edn). Penguin Books Ltd, Harmondsworth, UK.
Heavy Engineering Research Association. *Collected Papers on Quality Assurance*. HERA Information Centre, Auckland, New Zealand.
Hein, L., Andreasen, M.M. (1983). Integrated product development – a new reference system for methodical design. In: *Proceedings of ICED-83: WDK-10, Computer Aided Design & Design Methods*. Heurista, Zurich, pp 289–297.
Heskett, J. (2002). *Toothpicks and Logos – Design in Everyday Life*. Oxford University Press, Oxford.
Hollins, B., Pugh, S. (1990). *Successful Product Design: What to Do and When*. Butterworth, London.
Hollins, G., Hollins, B. (1991). *Total Design – Managing the Design Process in the Service Sector*. Pitman Publishing, London.
Houghton, A.F. (1986). Professional negligence in engineering practice. *Proceedings of the Institution of Mechanical Engineers* **200** (B-1), 33–36.
Hubbard, G. (1985). How to pick the personality for the job. *New Scientist* (31 January), 12–15.
Huber, P.W. (1988). *Liability: The Legal Revolution and its Consequences*. Basic Books, New York.
Hubka, V. (1982). *Principles of Engineering Design*. Butterworth Scientific, London.
Hudson Institute, Inc. (1999). *Mechanical Engineering in the 21st Century – Trends Impacting the Profession. A Report to the ASME Committee on Issues Identification*. ASME International, New York.
Institution of Production Engineers Working Party (1984). *A Guide to Design for Production*. Institution of Production Engineers, London.
Kagan, H.A., Van de Water, J. (1986). Design in jeopardy: the expanding legal responsibilities of engineers. *Journal of Professional Issues in Engineering* **112** (1), 58–67.
Kempner, T. (ed.) (1980). *A Handbook of Management* (3rd edn). Penguin Books Ltd, Harmondsworth, UK.
Kverneland, K.O. (1996). *Metric Standards for Worldwide Manufacturing* (2nd edn). ASME Press, ASME International, New York.
Leech, D.J. (1972). *Management of Engineering Design*. John Wiley & Sons, London.
Lipin, O.F. (1991). Gosstandart International: the USSR national system for standardization, metrology, and product quality control. *ASTM Standardization News* **19** (9), 44–47.
Lupton, T. (1983). *Management and the Social Sciences* (3rd edn). Penguin Books Ltd, Harmondsworth, UK.
Macdonald, R.M. (1985). Drawing up the purchasing specification. *Proceedings of the Institution Mechanical Engineers* **199** (B1).
McRobb, R.M. (1983). Customer-perceived quality levels. *Quality Assurance* **9** (4), 90–92.
Matousek, R. (1963). *Engineering Design – A Systematic Approach*. Blackie & Son Ltd, London.
Mintzberg, H. (1983). *Designing Effective Organizations (Structure in Fives)*. Prentice-Hall, Englewood Cliffs, NJ.
Muir, J. (1984). Incompetence at work. *CME (Chartered Mechanical Engineer)*, Institution of Mechanical Engineer (October), 48–49.
Muster, D., Mistree, M. (1984). Design in the systems age. *CME (Chartered Mechanical Engineer)*, Institution of Mechanical Engineer (February), 39–42.
Nadler, G. (1981). *The Planning and Design Approach*. John Wiley & Son, New York.
Nadler, G. (1986). Systems methodology and design. *Mechanical Engineering* **108** (9), 84–88.
Oakley, M. (1984). *Managing Product Design*. Weidenfeld & Nicholson, London.
Osgood, C.C. (1970). *Fatigue Design*. Wiley-Interscience, New York.
Ostrofsky, B. (1987). *Design, Planning and Development Methodology*. Prentice-Hall, Englewood Cliffs, NJ.
Pahl, G., Beitz, W. (1986). *Konstruktionslehre* (revised edn). Springer-Verlag, Berlin.
Parkinson, S.T. (1982). The role of the user in successful new product development. *R & D Management* **12** (3), 123–131.
Petroski, H. (1994). *Design Paradigms – Case Histories of Error and Judgment in Engineering*. Cambridge University Press, Cambridge.
Pitts, G. (1973). *Techniques in Engineering Design*. Butterworth, London.
Poynter, D. (1987). *Expert Witness Handbook: Tips and Techniques for the Litigation Consultant*. Para Publishing, Santa Barbara, CA.
Pugh, D.S. (ed.) (1984). *Organization Theory* (2nd edn). Penguin Books, Harmondsworth, UK.
Rothschild, E.E. (1987). *Product Development Management*. T. Wilson Publishing Company, Australia.
Rothwell, R., Schott, K., Gardiner, P. (1983). *Design and the Economy*. The Design Council, London.

Ryan, N.E. (ed.) (1988). *Taguchi Methods and QFD, Hows and Whys for Management.* American Supplier Institute (ASI), Dearborn, MI.

Sackleh, F.J. (1984). *Engineering and Leadership.* Paper 84-WA/Mgt-3. American Society of Mechanical Engineers, New York.

Samuel, A.E. (ed.) (1989). *Engineering Design and Manufacturing Management – Proceedings of Workshop on Engineering Design and Manufacturing Management, University of Melbourne, Australia, 21–23 November 1988.* Elsevier, Amsterdam.

Saunders, M. (1991). *ISO 9000 and Marketing in Europe: Should U.S. Manufacturers be Concerned?* U.S. Department of Commerce, National Institute of Standards and Technology, Gaithersburg, MD.

Sedig, K. (2003). *Swedish Innovations.* The Swedish Institute, Stockholm, Sweden.

Shigley, J.E. (1986). *Mechanical Engineering Design.* McGraw-Hill, New York.

Smith, C.O. (1983). Design requirements imposed by product liability. In: *Proceedings of ICED-83: WDK-10, Computer Aided Design & Design Methods.* HEURISTA, Zurich, pp. 265–269.

Strauss, J.B., Paine, C.E. (1937). *The Golden Gate Bridge: Report of the Chief Engineer.* Golden Gate Bridge Highway and Transportation District, San Francisco, CA. (50th Anniversary Edition 1987).

Sykes, R.N. (1980). *The Planning and Execution of Engineering Work in a Project Environment.* ASME Paper: 80-WA/Mgt-2. American Society of Mechanical Engineers, New York.

Takeuchi, H., Nonaka, I. (1986). The new product development game. *Harvard Business Review* (January–February), 137–146.

Anon. (1991). The need for ISO standards. *American Machinist* **135** (5), 27.

ASME. (1994). *The Why and How of Codes and Standards from the American Society of Mechanical Engineers.* ASME International, New York, 1994.

Topalian, A. (1980). *The Management of Design Projects.* Associated Press, London.

Toth, R.B. (1984). Putting the U.S. standards system into focus with the world. *ASTM Standardization News* (December), 16–20.

Tovey, M. (1984). Designing with both halves of the brain. *Design Studies* **5** (4), 219–228.

Turner, B.T. (1975). Creative approaches to engineering design. CME (Chartered Mechanical Engineer), Institution of Mechanical Engineer (November), 85–89.

Turner, B. (1982). Design audit. *Design Studies* **3** (3), 115.

VDI-Berichte 457 (1982). *Designers Reduce Manufacturing Costs: Methods and Techniques.* Verein Deutscher Ingenieure, Dusseldorf (201 pp).

VDI Guideline 2221 (1993). *Systematic Approach to the Design of Technical Systems and Products* (Translation/development of German edition 11/1986). Verein Deutscher Ingenieure, Dusseldor (44pp).

VDI 2223 (2004). Systematic embodiment design of technical products. Verein Deutscher Ingenieure, Dusseldorf (94pp).

VDI 2234 (1990). *Basic Economical Information for Design Engineers.* Verein Deutscher Ingenieure, Dusseldorf.

VDI 2235 (1982). *Economic Decisions in Design (VDI-Richtlinien).* Verein Deutscher Ingenieure, Dusseldorf (40 pp).

Wallace, K.M. (1988). Developments in design teaching in the Engineering Department at Cambridge University. *International Journal of Applied Engineering and Education* **4** (3), 207–210.

Wallace, K.M. (1990). A systematic approach to engineering design. In: Oakley, M. (ed.), *Design Management: A Handbook of Issues and Methods.* Basil Blackwell Ltd, Oxford, chapter 22.

Wesner, J.W., Hiatt, J.M., Trimble, D.C. (1995). *Winning with Quality – Applying Quality Principles in Product Development.* Addison-Wesley, Reading, MA.

Willis, T., Kaplan, M.P., Kane, M.B. (1989). Safety in design – an American experience. In: *Proceedings of the International Conference on Engineering Design, ICED-89.* Institution of Mechanical Engineers, London, paper C377/110.

White, N.A. (1983). Engineering management: the managerial tasks of engineers. *Proceedings of the Institution of Mechanical Engineers* **197B** (68), 1–13.

Wiele, L.E., Messner, M.E. (1984). Essential elements of a successful engineering and construction project. *Journal of Metals* (February), 41.

Wilson, B. (1984). *SYSTEMS: Concepts, Methodologies and Applications.* John Wiley & Sons, Chichester.

Index

ASME Boiler and Pressure Vessel Code, 62, 143, 163, 183, 186
Accounting systems, 39, 40
Approaches to managing engineering design
  holistic, 20
  quality engineering, 18, 149, 164
  systematic, 18–22, 63, 64, 119, 222
Appropriate technology, use of, 18, 32
Assembly
  layouts and models, 164
  storage and handling, 190, 191

Bids, see Project proposal, 95, 100
"Breakthrough" new products, 34
Brief, see Project proposal
  design, 68, 95–96
  safety, 229
British Standard BS 7000, 219
Budgets, see cost estimates

Cadillac, 205
Calculations, guidelines for, 159–160
Cambridge Engineering Selector, 62, 161
Cambridge University, vi, 3, 5, 98
Checklist
  conceptual design, 134
  design context, 44
  design specification, 116
  detail design, 198
  embodiment design, 170
  personnel profile, 88
  project profile, 76
  project proposal, 103
Clarification, see Task clarification
Clarity in design, 150
Communication channels and interactions, 64–65
Computers and engineering design, 20, 62

Concept, the
  developing, 141
  selection and evaluation of, 126
  as solution to the problem, 120, 122
Conceptual design
  checklist, 133–134
  divergent and convergent thinking during, 119–120
  drawings, 143
  estimating costs during, 127–129
  generating ideas during, 122–123
  models of, 120
  as a phase of design process, 25, 68–71, 234
  presentation guidelines for, 130–133
  selecting and evaluating concepts during, 120, 126
  a systematic approach to, 119–127
  vulnerability and weak spots of, 126
  work sheet, 135–136
Concurrent engineering, 18
Context
  checklist, 43–44
  engineering design, 1–2, 5, 9, 20, 23, 25, 28–29, 34, 51, 87, 233–234
  project, 3, 5, 20, 23, 25, 27–29, 31, 34–35, 43, 64, 233–234
  work sheet, 43, 45
Contracts, 95–97, 100–101
Convergent thinking, 119–121
Corporate influences, 29, 39
  accounting, 39
  benefits, 41
  management style, 39
  organizational behavior, 39
  structures, 39
  values, 41
Cost estimating
  recommended procedure for, 127

sample sheet of, 128
for testing and commissioning, 191
Costs, 66, 72–73, 97, 126–127, 164
competitive, 34
cumulative graphs of, 73
justification of, 34
Crescent® Wrench, 15
Customers
differences between users and, 37–38
expectations, 37–38, 205
see Users

Definition of design, 2
Demands and wishes, (also see
Requirements list), 110–111, 113–115
Design effort, 67–71, 197
and phase diagrams, 69
Design engineers, v
role of, v, 221
as distinct from inventors, 122
Design failures, 11–15, 220, 240
Design feedback, 20, 205
Design histories, 51
Design layout, 149, 164–165
Design manager, 58, 85
detail design and, 177, 178
as negotiator, 34
Design methods, 67, 68
in Britain, Germany and North
America, 18
Design problem, see Problem statement
Design process, 2
analysis of, 20
set in context with project, 23
models of, 18, 25, 29
four main phases of, 25, 68
"need" in, 95
outputs from, 25
phase overlap, 68
review and analysis of, 236
schematic of, 19
Design project
characteristic features of, 55
types of, 74
Design quality assessment work sheet,
211, 213
Design records, importance of, 109–110,
209
Design research, 20, 31, 40

Design review meetings, 51, 74, 115, 133,
169, 197, 211
Design specification, 111–114, 119
checklist, 116
Quality Function Deployment (QFD)
and, 107
timescale and, 109–110
work sheet, 117
Design standards and codes, see
Standards and codes
Design task, see Task Clarification
Design team, 58
activities of, 1, 67, 97, 233
building, 58
combination of design task and, 55
influences on, 83
managing, 83
motivating, 85
negotiating ability and power, 59
output from, 67
personal output from, 87
productivity of, 87
relationships, 58–59, 87
self-perception inventories of, 59
team roles in, 58
work effort of, 67
Design tools and techniques, 62, 66–67
Design, conceptual, see Conceptual
design
Design, detail, see Detail design
Design, embodiment, see Embodiment
design
Design, systematic approaches to, see
Systematic approaches to Design
Detail design, 25, 204
assembly during, 190
commissioning during, 191
deficiencies in leading to design
failures, 177
the design manager's role during, 178
drawings for, 189
as distinct from embodiment design,
68, 141
interaction of shape, materials and
manufacture during, 180
manufacturing drawings and
information produced during, 189
as a phase of design process, 25, 68
quality assurance and, 178

standard components and, 190
  testing during, 191
Divergent thinking, 119–120
Division of tasks, guidelines for, 156
Drawings, 18, 62, 100
  conceptual, 130
  detail, 189
  manufacturing, 189

EasyGreen® Lawn Spreader, 192–196
Economic loop for a typical project, 25–26
Effective engineering design, 43
  example of, 71
Efficient engineering design, 43
  example of, 71
Embodiment design, 25, 68, 141
  checklist, 170
  design standards and codes in, 162
  as distinct from detail design, 68, 141–149
  division of tasks during, 156
  guidelines for, 149–165
  incremental design approach and, 141
  using layouts and models during, 164
  as a phase of design process, 25, 68
  prototypes and testing in, 165
  work sheet, 171
Engineering analysis compared with engineering design, 108
Engineering design as compared with engineering analysis, 108
Engineering design process, see Design process
Engineering design team, see Design team
Engineering design, definition of, 2
  forensic analysis of issues, 235
Engineering project
  economic loop for, 25, 26
  typical inputs to and outputs from, 23–24
  environment, 25
Engineering, concurrent, 18
Enthusiasm, in design manager and design team, 39, 41, 85, 88, 130
Environmental issues, 31
Eschenbrenner v. Willson Safety Products, 207

European Committee for Electrotechnical Standardization (CENELEC), 224
European Committee for Standardization (CEN), 224
European Standards for Quality Management, 224
Examples, introduction to, 3

Fail-safe design, 152
Failure, see Design failures
Fisher & Paykel, DishDrawer®, 16
Force transmission, guidelines for, 155–158
Foreseeable misuse, 38, 206
Form design, 180–185
Formway Life chair project
  as an example, 5
  corporate influences on, 49
  design task influences on, 76–77
  design team influences on, 78–79
  design tools influences on, 79–80
  macroeconomic influences on, 46
  microeconomic influences on, 46, 48
  personnel influences on, 89–92
  example work sheets for, 47, 78, 90, 105, 118, 136, 172, 200, 214
Formway Furniture, Inc., 5, 197

Gantt chart, 99
Gasifier Test Rig (GTR), 3–5, 33, 37, 40–43, 61, 66, 68–74, 86, 98–99, 113–114, 127–128, 130–131, 165–168, 186–187
Generating ideas, 122
Guidelines for management of engineering design, vi, 1

Herald of Free Enterprise, capsize of, 158
Holistic approach to engineering design, 20
Hot melt coater/laminator machine, as example, 237
Humor, as a tool for design manager, 86

Incremental design approach, 24, 141
Influences
  corporate, 39
  on design task, 55, 97
  on design team, 58–60, 83

macroeconomic, 31
microeconomic, 34
Integrated product development, 18
International Standards for Quality Management, ISO 9000, 179
Intermediate technology, use of, 32
Inventor approach, v, 122
Involvement, by design manager, 41

Jaguar, 87, 164
Japanese companies
 incremental design approach in, 24, 141
 Quality Function Deployment (QFD) in, 63, 179
 Taguchi Methods in, 63, 179

Knoll Inc., 7, 63, 197

Layouts
 use of, 164
 guidelines for, 165
Levels of resolution, 20–21, 23, 29
 corporate, 29
 macroeconomic, 31
 microeconomic, 34
 personal or design team personnel, 25, 83
 project, 55
Life Chair, 5, 46–51, 63, 76–80, 84, 88–92, 101–102, 105, 110, 118, 120, 123–124, 127, 133, 136–140, 169, 172–175, 188–189, 197, 200–203, 214
Life-cycle engineering, 18, 19
Life-cycle, see Product life-cycle
Litigation, and design records, 208
Low Cost Rotary Lawn Spreader, see "Scotts" EasyGreen® lawn spreader

Macroeconomic influences, 31
Maintenance of products in service, 208
Management and managing engineering design, 39–43
 improving within a company, 17, 75
 skills for, 42
 structural approaches to, 39
 style of, 42
 tips for, 22, 51, 80–81, 92, 102, 115, 140, 176, 204, 212
 phases for, 68
Management, upper, 39, 100

Manager, design, see Design manager
Manufacturing, 20
 drawings, 189
 information, 189
 interaction with materials and shape, 184–185
 schematic of basic design and manufacturing process, 19
Manz Engineering Ltd, electronic deadbolt example, 179
Market and marketing, 18, 34
Mass-produced products, 3, 63, 182
Materials
 Cambridge Engineering Selector and, 62, 161–162
 guidelines for selection of, 161
 interaction with shape and manufacture, 180–185
 specification, 111
 tramp metal in steel, 181
McKinsey 7-S Framework, 39
Methods, design, see Design methods
Microeconomic influences, 34
Millennium Footbridge, 11
Models, use of, 164
Monitoring
 design process, 67–74
 design team, 83–88
Motivation, in design team, 85

Need, in design process, 18, 34
Negotiating, 59
 design manager's role in, 34, 100
 team role in, 59

O. M. Scott and Sons Company, 192
Organizational behavior, 39
Outputs
 from design team, 67
 from phases, 20

Percent completion, 68–74
 phase diagrams as a means to assess, 68–70
Personnel profile
 checklist, 88
 work sheet, 88–89
Phase diagrams, 68–74
 percent completion as shown by, 68, 70, 73

timescale estimates based on, 73
Phases of design process
  general, 68
  conceptual design, 119
  detail design, 177
  embodiment design, 141
  proposal, 95
  task clarification, 107
Problem statement, 107
Problem, defining the, 120
Process, see Design process
Product design specification, see Design specification
Product integrity board meetings, 75
Product liability, 206–209
  lawsuit concerning Willson headphones, 207
  litigation and, 208–209
  standards and codes and, 163, 209
  use and abuse and, 206
Product life-cycle, 34
Product planning, 34
Project, engineering design
  characteristic features of, 55, 191
  context, 23
  economic loop for, 26
  plan using Gantt chart, 110
Project brief (also see Project proposal), 95
Project profile
  checklist, 76–77
  work sheet, 76
Project proposal
  bids and, 95
  checklist, 101, 103
  debriefing on, 101
  guidelines for preparing, 96–98
  negotiations and oral presentations for, 100
  project brief and, 95
  request for, 95
  structure and content of, 97
  work sheet, 101, 104
Proposal, (also see Project proposal)
Prototypes, 165
Purchased components
  and mail-order catalogs, 164

Quality Function Deployment (QFD), 63, 107, 179

product design specification and, 107, 111
Quality assurance, 20, 179
Quality management, 75, 179, 219, 222

Random influences, 33
Records, 209
  for future reference, 2
  importance of keeping, 109, 209
Redundant design, 152
Relationships in design team, 87
Request for proposal, 95, 97
Requirements list, 110
  demands and wishes and, 110
  design specification and, 111
Research, design, 3, 5, 31
Resolution level, see Levels of resolution
Resources, 34, 36
Review meetings, 74, 169, 197
Risk, technical and financial, 57, 97, 209

Safe-life design, 152
Safety
  direct and indirect approaches to in design, 152–153
  definitions of, 220
  embodiment design and, 151
  hierarchy for design, 151
  personal protection, 153
  product liability and, 206
  standards and codes and, 162–163
  training and instructions, 153
  use and abuse and, 206
  warnings, 153
Scott Fetzer
  product integrity board meetings of, 75
  quality assurance by, 178–179
"*Scotts*" *EasyGreen*® lawn spreader, 192–198; (also see *Easy Green*® lawn spreader)
Segway™ Human Transporter, 15–17
Selecting and evaluating concepts, 126
Self-help, guidelines for, 157
Self-Perception inventories, 59–61
Shape
  form design and, 184
  interaction with materials and manufacture, 180–185

Shared values, 41
Simplicity in design, 150
Simultaneous engineering, 18
Sony Walkman, management style and, 42
Space Shuttle Challenger, 22, 59, 151, 154, 159
   as an example using design quality assessment work sheet, 211, 215
   report of presidential commission on, 211
Specification (also see Design specification and Materials specification), 111–114
Stability in design, guidelines for, 158
Standard components, 190
Standards and codes, 162–163, 217
   catalogs of, 223
   contacts for, 223–231
   definitions of, 218
   European, 224–227
   guidelines for, 162–164
   international, 221, 222, 223–231
   national, 221, 222
   quality management, 219
   reference papers on, 221
   safety, 220–221
   use of, 162–163
Structured approach to managing engineering design, 29, 209
Summary, 233
Systematic approach to engineering design, 22, 36, 100, 119, 141, 178, 209

Taguchi Methods, 63, 179
   in perspective with overall design process, 64
Task clarification phase, 25, 107
Task specification, 111
Team, see Design Team
Technological cycles, 32
Tenacity, as factor in design projects, 85, 88
Testing, 165
   commissioning and, 191
   prototypes and, 165
   Underwriters Laboratory (UL), 191, 230

Texas A&M University, stadium example, 109
Timescale, project, 57, 99
   estimating, 70, 109
   phase diagrams and, 68–74
Tips for Management, 22, 51, 80–81, 92, 102, 115, 140, 176, 204, 212
Tolerances, dimensional and geometric, 184, 189
Tooling, 181, 184
Tri-axis transfer press, as example, 57–58, 160
Triodyne Inc., 221, 229
Triodyne Safety Information Center, vi

Underwriters Laboratory (UL), 191, 230
Users, 37
   differences between customers and, 37
   expectations of, 205
   foreseeable misuse by, 206
   litigation by, 206
   maintenance of products in service by, 208
   needs of, 34, 37
   satisfaction of, 38

Viewpoint
   concept of, 20
   engineering design, 23, 25

Willson headphones example, 207
Windowing process, 25, 29
Wobbly bridge syndrome, 11
Work effort, see Design effort
Work sheet
   conceptual design, 135
   design context, 45
   design quality assessment, 213
   design specification, 117
   detail design, 199
   embodiment design, 171
   personnel profile, 89
   project profile, 77
   project proposal, 104
Work statement, 95
World Trade Center, collapse of towers, 13–15